CALCULUS with the TI-89

Brendan Kelly

Brendan Kelly Publishing Inc.
2122 Highview Drive
Burlington, Ontario L7R 3X4
Canada
www.brendankellypublishing.com

ISBN-13: 978-1-895997-13-2
ISBN-10: 1-895997-13-5

ATTENTION EDUCATIONAL ORGANIZATIONS

Quantity discounts are available on bulk purchases of this book for educational purposes or fund raising. For information, please contact:

Brendan Kelly Publishing Inc.
2122 Highview Drive
Burlington, Ontario
L7R 3X4

Telephone: (905) 335-3359 Fax: (905) 335-5104

INTRODUCTION

WHY IS THE GRAPHING CALCULATOR IMPORTANT?

The widespread acceptance of the graphing calculator as an important tool for the teaching and learning of mathematics is now evolving into a compelling assertion of its necessity. Mathematicians and mathematics educators recognize that this new technology, though not a panacea, is an essential mathematical tool which is changing irrevocably the content and pedagogy of high school and college mathematics.

WHY IS THIS BOOK NEEDED?

It is not feasible to expect students to master all the functions and menus of the TI-89 before applying it to the study of calculus. On the contrary, it is important that instruction in mathematics draw upon the calculator functions as needed, neglecting those aspects which are not fundamental to the mathematical development. Achieving this integration of machine and mathematics requires a sequence of carefully designed instructional activities that showcase a technology-conscious approach to each mathematical concept. This book is designed to be such a resource.

TO THE STUDENT

Your TI-89 calculator and this book will greatly enhance your study of calculus by helping you to explore important skills and concepts. Textbook treatments of calculus are typically formal and abstract. Their quest for rigor often obscures the powerful ideas that propelled the early development of this beautiful subject.
As you work through the activities of this book, you will compute the instantaneous velocity of a space shuttle, calculate the acceleration of the Viper RT/10, explore the trajectory of baseball's longest home run and graph a surface in 3-D with finite volume and infinite surface area. We hope you enjoy your journey through the concepts, content and cartoons.

TO THE TEACHER

In writing this book, we have sought to develop the main skills and concepts of elementary calculus with the vitality and vigor embodied in the notion of change itself. Formal definitions and proofs that usually bury students in needless details have been downplayed in favor of heuristic explorations of the concepts of limit, slope, and area. The solutions to problems are usually modeled with and without the TI-89 to enable students to solve problems in various ways. This feature also facilitates instruction in classes in which only a subset of the students have access to the TI-89. For students who require further facility with the algebra or statistics functions of the TI-89, refer to the other publications listed on page 96. We hope this resource will help you in your important teaching endeavor.

Table Of Contents

Unit 1 — Limits of Sequences, Series & Functions

Exploration 1	Limits of Sequences	8
Exploration 2	Limits of Series: The Quest to Capture the Infinite	12
Exploration 3	Areas as Limits of Series: The Method of Exhaustion	16
Exploration 4	Limits of Functions & Asymptotic Behavior	20

Unit 2 — The Concept of the Derivative

Exploration 5	Distance, Rate & Time	26
Exploration 6	The Derivative as a Limit	30
Exploration 7	Functions and their Extrema	34

Unit 3 — Formal Differentiation

Exploration 8	Derivatives of Polynomials & Rational Functions	38
Exploration 9	Derivatives of Algebraic Functions	42
Exploration 10	Derivatives of Exponential Functions	46
Exploration 11	Derivatives of Logarithmic Functions	49
Exploration 12	Derivatives of Trigonometric Functions	53
Exploration 13	The Chain Rule & Related Rates	57

Table Of Contents

The Integral — Concept & Applications

Exploration 14	The Fundamental Theorem of Calculus	62
Exploration 15	Integrals of Exponential, Log, & Trig Functions	66
Exploration 16	Integration Techniques	70
Exploration 17	Applications of Integration in 2-Dimensions	74
Exploration 18	Applications of Integration in 3-Dimensions	77

Selected Solutions 81

Non Sequitur

This book attempts to teach mathematics and the use of the TI-89 Graphics calculator at the same time. To do this effectively, it has been necessary to simulate the calculator keys and fonts on the TI-89 and to generate screen displays. Achieving these elements required early access to beta versions of the TI-89 and the TI-89 Link. We are indebted to Len Catleugh and Tom Ferrio of *Texas Instruments* whose continuing help and support in providing early versions of software and hardware have made these publications possible.

We acknowledge our debt to David Bernklau of Stuyvesant High School in New York City for his painstaking editing of earlier versions of this manuscript. In removing anomalies and identifying omissions that escaped the attention of the author, he has significantly enhanced the final presentation of the material. My wife, Teri, has also contributed significantly in purging typos and suggesting enhancements to various editions of the manuscript. Her continuing support is a critical component in the preparation of these instructional materials.

For a resource to be truly effective in the classroom, it must be free of ambiguities. Instruction must be clear and activities must be sequenced to facilitate student learning. The only way to achieve this is to conduct extensive field testing. We are indebted to the following students enrolled during the 1996-97 academic year at OISE/UT, University of Toronto, for working through draft versions of the manuscript and providing innumerable improvements.

Elaine Alexiou	Ian Bain	Kathy Belevski	Matt Brown
Sonya Kim	Albert Koöij	Karen Leckie	John Pascual
Caralee Paul	Anita Princiotto	Noor Rehemtulla	Marie Shafi
Claude Sulpizi	Nicole Tanfara	Robyn Temkin	Henry Van Bemmel
Marsha Wieringa	Susan Woolam		

Encouragement from those of you who use our books for the TI-92 has inspired us to modify that trilogy for the TI-89. Since the TI-89 and the TI-92 calculators have almost identical capabilities, this book develops the same mathematical content as *Investigating Calculus with the TI-92*. It was necessary to change only the keying sequences and screen displays to match the TI-89. (The parallel structure and content of the TI-89 and TI-92 books will facilitate the instruction in classes that have partial class sets of both types of calculators.) We appreciate your kind words and will continue to publish materials to facilitate your use of this wonderful technology in the exploration of mathematical ideas.

Limits of Series & Functions

"'C' IN ASTROPHYSICS, 'B MINUS' IN CALCULUS... WHAT KIND OF GENIUS ARE YOU?"

Mathematical Concepts

- arithmetic sequences
- geometric sequences
- Fibonacci sequence
- defining sequences by recursion
- graphs of sequences
  - web plots
  - time plots
- the logistic equation
- finite and infinite series
- sums of arithmetic & geometric series
- Sigma notation
- method of exhaustion
- Zeno paradoxes
- the concept of the limit
- asymptotic behavior

TI-89 Commands

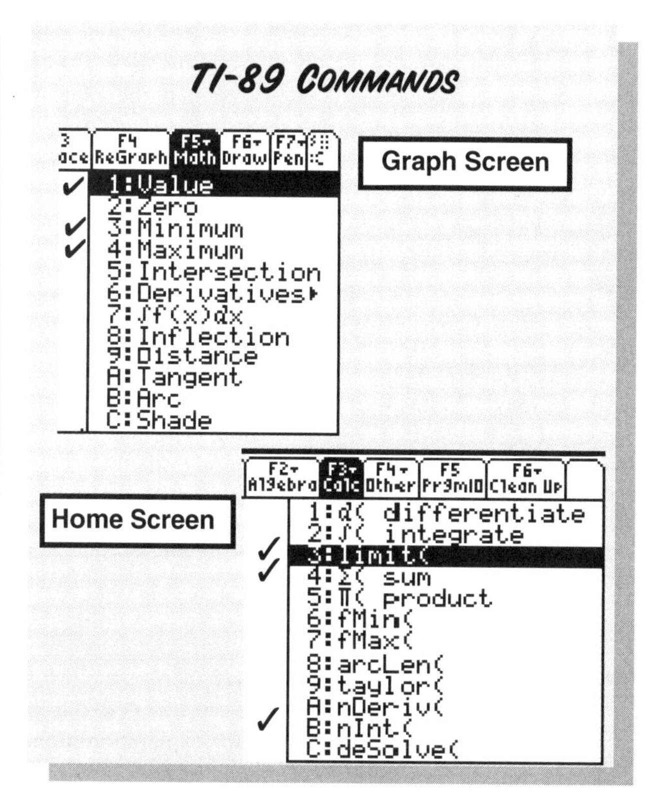

Exploration 1 — Limits of Sequences

Smith Coll., Rare Book and Manuscript Library, Columbia University

Leonardo da Pisa (Fibonacci) c. 1180 - 1250

Leonardo da Pisa, considered by some scholars to have been the greatest mathematician of the Middle Ages, published his famous treatise, *Liber abaci* (book of the abacus) in 1202. In this book, Leonardo da Pisa, better known as Fibonacci, presented the following problem:

> How many pairs of rabbits will be produced in a year, beginning with a single pair if in every month each pair bears a new pair which becomes productive exactly two months after birth?

By drawing a tree diagram showing the number of pairs of rabbits at the end of each month, we discover that the number of pairs of rabbits in the first six months are given by the following sequence, 1, 1, 2, 3, 5, 8, ... This famous sequence, named the *Fibonacci* sequence in honor of Leonardo da Pisa has the property that each term is the sum of the two previous terms.

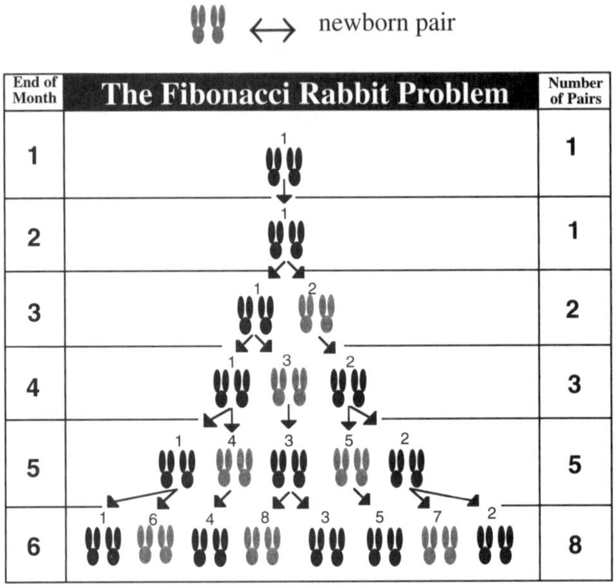

In the tree diagram on the right, we mark the first pair of rabbits with the number 1. Since they do not produce offspring in the first 2 months, there is only one pair at the end of two months. On the first day of the third month, they give birth to another pair (labelled 2) so there are two pair at the end of the third month. The arrows in the tree diagram show the offspring in the successive months up to the end of the sixth month.

↔ newborn pair

End of Month	The Fibonacci Rabbit Problem	Number of Pairs
1	1	1
2	1	1
3	1 2	2
4	1 3 2	3
5	1 4 3 5 2	5
6	1 6 4 8 3 5 7 2	8

The Fibonacci sequence is one of an infinite number of possible sequences. To define a sequence, we merely assign to each positive integer n a particular number $u(n)$. The number $u(1)$ is taken to be the first term of the sequence, $u(2)$ is the second term of the sequence, ... and $u(n)$ is the nth term. If we define $u(n)$ as an algebraic expression, then we have defined all the terms of the sequence. In effect, a sequence is a function defined on the positive integers (rather than on the real numbers). In general,

> **Definition:** Any function u defined for all positive integers from 1 to N is said to define a *sequence of N terms*, and $u(n)$ is said to be the nth *term* of the sequence u.

If a sequence has an infinite number of terms, we call it an *infinite sequence*.

Arithmetic Sequences

The first five terms of the sequence $u(n) = 5 + 3(n - 1)$ are: 5, 8, 11, 14, and 17.
Each term is 3 greater than the previous term. That is, $u(n) - u(n-1) = 3$. Such a sequence in which successive terms differ by a constant is called an *arithmetic sequence*. This constant is called the *common difference*.

The first few terms of a general arithmetic sequence with first term a and common difference d are given by:

$$a,\ a + d,\ a + 2d,\ \dots\ \boxed{a + (n-1)d,}\ \dots$$

common difference d (pointing to $a + d$); n^{th} term $u(n)$ (pointing to $a + (n-1)d$)

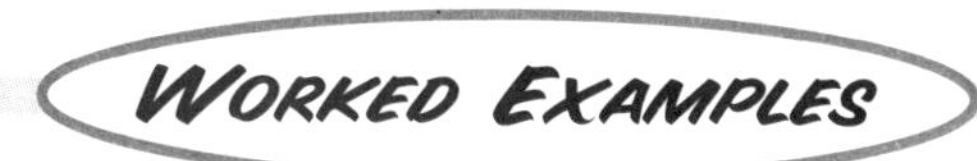

Geometric Sequences

The first five terms of the sequence defined by $u(n) = 2^n$ are: 2, 4, 8, 16 and 32.
Each term in the sequence is double the previous term. That is, $u(n) = 2u(n-1)$ Such a sequence in which successive terms are in a fixed ratio is called a *geometric sequence*. The fixed ratio is called the *common ratio*. The first few terms of a general geometric sequence with first term, a and common ratio, r are given by:

$$a, \quad ar, \quad ar^2, \ \ldots, \ \boxed{ar^{n-1}}, \ldots$$

↑ common ratio r ↑ n^{th} term $u(n)$

CAN YOU DETERMINE WHICH OF THESE SERIES ARE ARITHMETIC, WHICH ARE GEOMETRIC AND WHICH ARE NEITHER?

WORKED EXAMPLE 1

a) List the first 10 terms of each of the following sequences and evaluate the 20th term.

(*i*) $u1(n)=4-7(n-1)$ (*ii*) $u2(n)=2^n$ (*iii*) $u3(n)=\left(\frac{3}{4}\right)^n$ (*iv*) $u4(n)=3n^2+2n-1$ (*v*) $u5(n)=\left(1+\frac{1}{n}\right)^n$

b) For each sequence in part a) what is the value of $u(n)$ as n approaches ∞?

SOLUTION

a) To obtain the first 10 terms of each sequence, we could substitute the values $n = 1, 2, 3, \ldots 10$ into each expression for $u(n)$. To obtain the 20th term, we can substitute $n = 20$ into each expression for $u(n)$.

(i) $u1(20) = 4 - 7(19)$ or -129 (ii) $u2(20) = 2^{20}$ or 1 048 576 (iii) $u3(20) = 0.0031712\ldots$
(iv) $u4(20) = 3(20)^2 + 2(20) - 1$ or 1239 (v) $u5(20) = (1 + 1/20)^{20}$ or $2.65329\ldots$

Alternatively, we can use the **seq(** command to display the first 10 terms on the Home screen.
To access the **seq(** command, we press: **2nd** [MATH] **3** **ENTER**.
Then we enter on the Home screen, the commands shown on the command line and press **ENTER**.

The first 5 terms of sequence $u1$ are displayed.
To obtain the next five terms, we highlight (press ▲) the terms of $u1$ and scroll right as shown in the display.

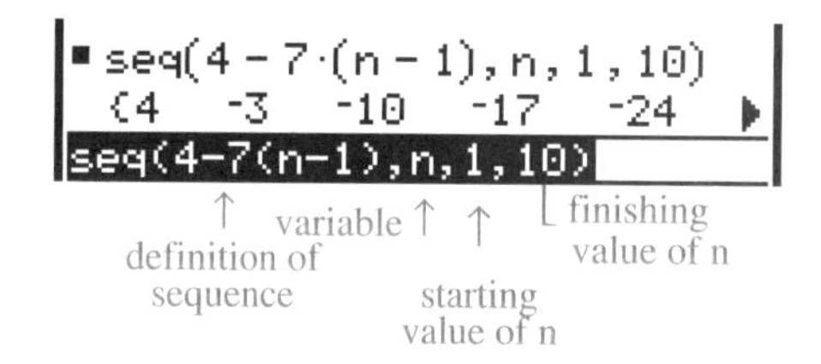


```
■ seq(4 - 7·(n - 1), n, 1, 10)
 {-31  -38  -45  -52  -59}
seq(4-7(n-1),n,1,10)
```

We obtain: $\{4, -3, -10, -17, -24, -31, -38, -45, -52, -59\}$.

To obtain $u1(20)$, we merely change the starting and finishing values of n on the command line to 20. This yields $u1(20) = -129$ as shown in the display.

```
■ seq(4 - 7·(n - 1), n, 20, 20)
                          {-129}
seq(4-7(n-1),n,20,20)
MAIN        RAD AUTO      FUNC      1/70
```

Proceeding as above for $u2$, $u3$, $u4$, and $u5$, we find: Note: To express a fraction as a decimal press:

$u2(n) = 2, 4, 8, 16, 32, 64, 128, 256, 512, 1024, \ldots$ and $u2(20) = 1\,048\,576$
$u3(n) = 0.75, 0.5625, 0.4219, 0.3164, 0.2373, 0.1780, 0.1335, 0.1001, 0.0751, 0.0563$ and $u3(20) = 0.0031712\ldots$.
$u4(n) = 4, 15, 32, 55, 84, 119, 160, 207, 260, 319$ and $u4(20) = 1239$
$u5(n) = 2, 2.25, 2.37, 2.441, 2.488, 2.522, 2.546, 2.566, 2.581, 2.594$ and $u5(20) = 2.65329\ldots$

b) The value that $u1(n)$ approaches as n approaches ∞ is denoted $\lim_{n\to\infty} u1(n)$.← Read "The limit of $u(n)$ as n approaches infinity."

To evaluate this limit, press: **F3** **3**. Enter $u1(n)$, n, ∞ as shown in the display. This yields $\lim_{n\to\infty} u1(n) = -\infty$.

Similarly we find the limits for $u2(n), \ldots, u5(n)$ are respectively, ∞, 0, ∞ and e. ← What's this???

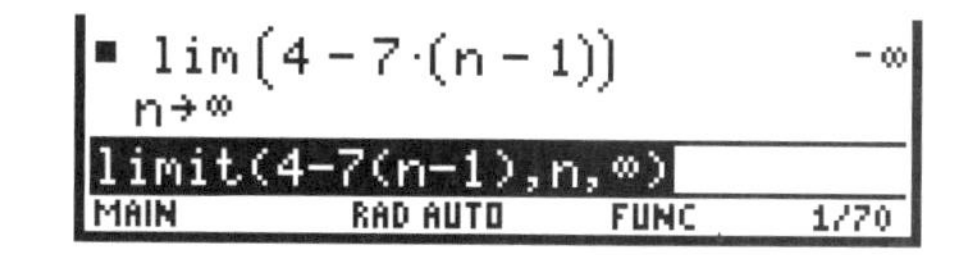

WORKED EXAMPLES

We observed that the Fibonacci sequence was defined by relating the nth term $u(n)$ to the two preceding terms, $u(n-1)$ and $u(n-2)$. Also we saw how the arithmetic and geometric sequences could be defined by relating the nth term $u(n)$ to the preceding term $u(n-1)$. Sequences which are defined by relating the nth term to one or more previous terms are said to be defined *by recursion*. Such sequences are very important in modeling dynamic systems such as population growth, where the population of the nth generation is a function of the population of the $(n-1)^{th}$ generation.

"This must be Fibonacci's."

WORKED EXAMPLE 2

a) Create a table to display the first 8 terms of each sequence.

i) $u1(n) = u1(n-1) - 7$

ii) $u2(n) = 2u2(n-1)$

iii) $u3(n) = u3(n-1) + 2\sqrt{u3(n-1)} + 1$

iv) $u4(n) = \dfrac{1}{u4(n-1)} + 1$

v) $u5(n) = u5(n\text{-}1) + u5(n\text{-}2)$

b) Graph the first 21 terms of sequence $u4$. Then trace to evaluate its 20th term.

SOLUTION

a) Before defining each sequence, we press: MODE ▶ 4 ENTER to obtain SEQUENCE mode. Then we press: ◆ [Y =].
Then $u1$, $u2$, …, $u5$ are defined as shown in the display on the right. We must enter the starting values, $ui1$, $ui2$, …, $ui5$, because each term of the sequence is determined by referring to the previous term(s).
To start the table at $n = 1$, we press: ◆ [TblSet] and complete the dialog box by entering 1 for tblStart and 1 for Δtbl.

Use F4 to select ✓or de-select a sequence.

```
F1▾   F2▾  F3   F4 F5▾ F6▾   F7
Tools Zoom Edit ✓  All Style Axes...
▴PLOTS
✓ u1=u1(n - 1) - 7
  ui1=-7
✓ u2=2·u2(n - 1)
  ui2=2
✓ u3=u3(n - 1) + 2·√u3(n - 1) ▸
  ui3=1
          1
✓ u4=  ―――――――― + 1
       u4(n - 1)
  ui4=1
✓ u5=u5(n - 1) + u5(n - 2)
  ui5={1  1}
ui5={1,1}
MAIN        RAD AUTO       SEQ
```

Press ENTER . To create the table, we press: 

n	u1	u2	u3
1.	-7.	2.	1.
2.	-14.	4.	4.
3.	-21.	8.	9.
4.	-28.	16.	16.
5.	-35.	32.	25.

u3(n)=25.
MAIN RAD AUTO SEQ

To view u4 and u5, we press the 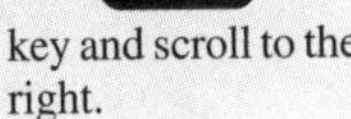 key and scroll to the right.

b) To graph the first 21terms of $u4$, press: ◆ [WINDOW],
Enter the values shown in this display.
Press ENTER, then ◆ [GRAPH]

```
nmin=1.
nmax=21.
plotStrt=1.
plotStep=1.
xmin=-1.
xmax=21.
xscl=1.
ymin=-1.
ymax=
MAIN        RAD A
```

We then trace (F3) along $u4$ to obtain $u4(20)$ = 1.618034…, as in the display. ↑ Hmmm… Look familiar??

The 4 indicates that the cursor is on $u4$.

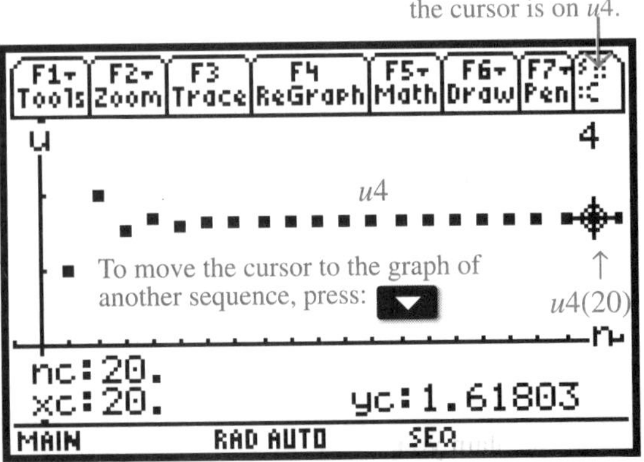

Exercises & Investigations

❶. Write a sentence to define each of the following terms. Give an example of each.
a) a finite sequence
b) an infinite sequence
c) an arithmetic sequence
d) a geometric sequence

❷. If $u(n)$ defines an infinite sequence and L is a real number, explain what is meant by $\lim_{n\to\infty} u(n)=L$.

❸. Indicate whether the following statement is true or false:
An infinite sequence is a function defined on the positive integers.
Justify your answer.

❹. List the first 10 terms of each sequence.
a) $u1(n) = 95 - 9(n-1)$
b) $u2(n) = 3^{n-1}$
c) $u3(n) = 8 - 2^n$
d) $u4(n) = 2^{-n}$
e) $u5(n)=\left(1+\frac{2}{n}\right)^n$

Identify each sequence as an arithmetic sequence, geometric sequence, or neither type.

❺. Calculate the 18th term of each sequence in exercise ❹.

❻. a) Graph the first 20 terms of sequences $u4(n) = 2^{-n}$ and $u5(n)=\left(1+\frac{2}{n}\right)^n$ in the window: $0 \le x \le 21$ and $-1 \le y \le 10$.
b) Trace along each graph in a) to evaluate the 20th term of each sequence.
c) Evaluate $\lim_{n\to\infty} u4(n)$ and $\lim_{n\to\infty} u5(n)$.

❼. List the first 15 terms of each sequence.

a) $u(n)=\sqrt{u(n-1)}$
first term: 2

b) $u(n)=u(n-1)-u(n-1)^2$
first term: 0.75

c) $u(n)=\cos u(n-1)$
first term: 0

d) $u(n)=2.8u(n-1)[1-u(n-1)]$
first term: 0.6

❽. Evaluate $\lim_{n\to\infty} u(n)$ for each sequence in exercise ❼.

❾. Draw the next row in the Fibonacci rabbit tree diagram on page 8 to verify that there are 13 rabbits at the end of the seventh month.

❿. Compare the sequences in *Worked Examples* 1 and 2. Find a pair of sequences which are the same. Explain why the two different definitions generate the same sequence. Can every arithmetic sequence be written in both function and recursive form? Is this true also for every geometric series? Explain your answer.

⓫. In 1845, P. F. Verhulst formulated a law for population growth expressed by the *logistic* equation,

$$u(n) = \lambda u(n-1)[1-u(n-1)]$$

where $u(n)$ denotes the population of the nth generation and λ is a constant associated with a particular species.

a) Assuming an initial population of $u(0) = 0.6$ units, graph the populations of the first 40 generations (i.e. $u(n)$ vs. n) for each of these values of λ:
(i) $\lambda = 1.6$ (ii) $\lambda = 3.1$ (iii) $\lambda = 3.5$

Hint: Set the window variables as shown in this display by pressing [◆] [WINDOW]

```
nmin=1.
nmax=40.
plotstrt=1.
plotstep=1.
xmin=-1.
xmax=45.
xscl=1.
ymin=-.2
ymax=1.
yscl=1.
```

For each value of λ, trace along your graph to find the population of the 40th generation; i.e., $u(40)$.

b) For each value of λ, indicate whether there is a limiting "stable value" toward which the population converges. That is, for each value of λ, indicate whether there is some limiting population L such that $\lim_{n\to\infty} u(n)=L$. Justify your answer by interpreting your graphs.

c) To compare the nth term of a sequence with the $n-1^{th}$ term for large n, mathematicians have developed the so-called *web plot* (because it resembles a spider web). The web plot is a scatterplot of the points $(u(n-1), u(n))$.

If the sequence $u(n)$ approaches a limit for large n, then the points in the web plot will lie on or close to the line $y = x$. Any point $(u(n-1), u(n))$ on a web plot projects horizontally onto the point $(u(n), u(n))$ on the line $y = x$. Why? The point $(u(n), u(n))$ on the line $y = x$, projects vertically onto the point $(u(n), u(n+1))$ on the web plot. Why? Therefore, starting at any point on the web plot, with vertical coordinate $u(n)$ and projecting horizontally onto $y = x$ and then vertically onto the web plot, maps us onto a point with vertical coordinate $u(n+1)$. This provides a visual map of the long term behavior of $u(n)$.

To create web plots of the three logistic sequences above, press: [◆] [Y =] [2nd] [F7] and select WEB, opposite Axes, and TRACE opposite Build Web. Then graph the sequences using these window settings and trace.

Write a brief report to describe and interpret your web plots.

Zeno's Paradox

Almost 2500 years ago, the ancient Greek philosopher Zeno of Elea (495 – 435 B.C.) presented several paradoxes which baffled the people of his day. One of the most famous of these paradoxes is called the *Dichotomy*. The Dichotomy paradox argues that a runner cannot travel a distance of one stadium (an ancient Greek unit of distance) because he must first travel half that distance; i.e., 1/2 of a stadium, but before he can travel the remaining distance of 1/2 stadium, he must travel half of this distance; i.e., 1/4 stadium, and after this, he must travel 1/8 stadium… and so on. Therefore, the total distance (in stadia) the runner must travel before he can traverse 1 stadium is:

$$\frac{1}{2}+\frac{1}{4}+\frac{1}{8}+\ \dots\ +\frac{1}{2^n}+\ \dots$$

Zeno argued that this distance is the sum of an infinite number of finite quantities, and is therefore infinite. Since the runner cannot travel an infinite distance in a finite time, he cannot travel a distance of one stadium.

The Pythagoreans had assumed that distances and time intervals were magnitudes that could be represented by discrete pebbles called *calculi* (from which our word *calculus* is derived). The dichotomy paradox challenged this discrete model of time and distance. It took another 2000 years for humans to discover how to deal with *infinite series*, such as the one shown above, in a rigorous way.

AT THIS RATE, I'LL NEVER GET TO THE STADIUM!

1 1/2 1/4 0

One way to sum the infinite series given above, is to imagine that you have a rod of pure gold that is infinitely divisible and measures one meter in length. You begin by dividing the rod into two equal parts and placing one part into a safe deposit box. You then break the remaining half into two equal parts, and place one of these two parts into the safe deposit box. What is the sum of the lengths of all the parts in the safe deposit box after you have done this n times? The sum of the lengths of the pieces in meters (called the nth *partial sum* and denoted $S(n)$) is:

$$S(n)=\frac{1}{2}+\frac{1}{4}+\frac{1}{8}+\ \dots\ +\frac{1}{2^n}$$

No matter how often we repeat this process, the total length of the pieces in the safe deposit box can never be more than 1 meter. Why? Furthermore, by repeating the process a large number of times, we can achieve a total length as close to 1 meter as we choose. (How do we know this?) Therefore, the infinite sum given above is larger than any number that is less than 1, yet it cannot exceed 1, so it must be equal to 1. We write:

$$\begin{aligned}\frac{1}{2}+\frac{1}{4}+\frac{1}{8}+\ \dots\ +\frac{1}{2^n}+\ \dots &= \lim_{n\to\infty} S(n)\\ &= \lim_{n\to\infty}\left(\frac{1}{2}+\frac{1}{4}+\frac{1}{8}+\dots+\frac{1}{2^n}\right)\\ &= 1\end{aligned}$$

This constructive method of generating the infinite series from a finite length, shows that it is possible, in some sense, for an infinite number of finite quantities to have a finite sum. We define the sum represented by an infinite series to be the limit of the sequence of partial sums, $S(1), S(2), S(3),\dots, S(n),\dots$ as $n\to\infty$.

Note: Strictly speaking, a series is a sum, so it is redundant to speak of the "sum of a series". However, we will use this phrase to mean "the sum of the terms of the corresponding sequence".

Worked Examples

According to an old legend[1], chess was invented over a thousand years ago by the servant of a powerful Hindu rajah. The rajah was so delighted with this invention, that he offered the servant any reward of his choosing. The modest servant requested only that he receive a grain of wheat for the first square of the chess board, 2 grains of wheat for the second square, 4 grains for the third square, and so on, with each square containing double the number of grains on the previous square. When the rajah discovered he could not fulfill the servant's demand, he ordered his execution!

Why was the rajah unable to fulfill the servant's demand?

Worked Example 1

Determine how many grains of wheat would be required to cover the entire chessboard.

Solution

Since there are 64 squares on a chessboard, we are required to find the sum of the series

$$1 + 2 + 4 + 8 + 16 + \ldots + 2^{63}.$$

We can sum this series directly using the **sum(** command together with the **seq(** command. We can enter these commands from the keyboard, or alternatively access these commands from the MATH menu by pressing these keys:

To define the series we wish to sum, we complete the entry as shown in the display.

sum(seq(2ⁿ, n, 1, 63, 1)

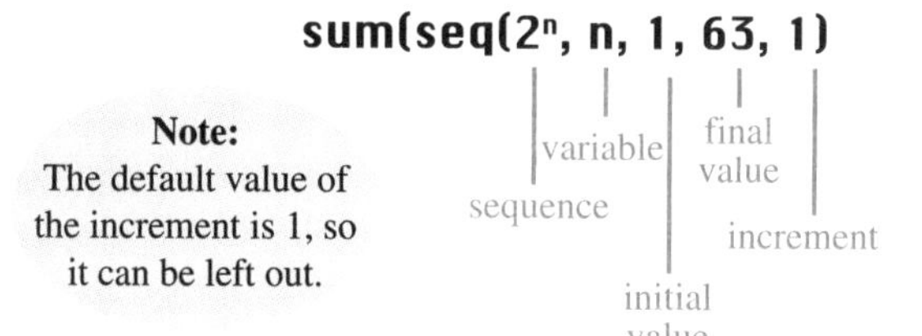

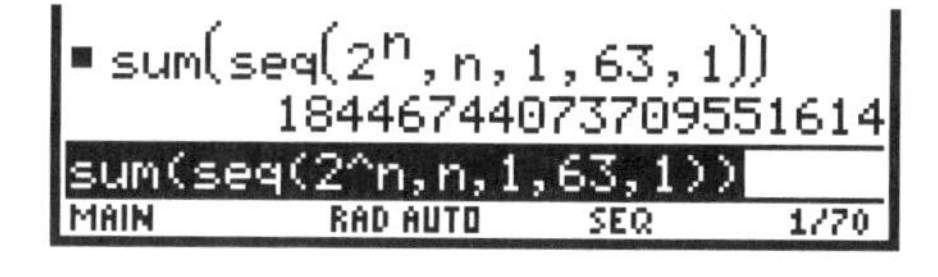

The display shows that the sum of this series is 18,446,744,073,709,551,614. That's a lot of wheat!

Σ Sigma Notation

Mathematicians have developed an abbreviated notation for the **sum(seq(** construction (used in *Worked Example* 1) called *sigma notation*. The notation is easily understood by studying this example.

We see: $3^2 + 4^2 + 5^2 + \ldots + 103^2$

We say" The sum of the sequence n^2 as n runs from 3 to 103."

We write:

$$\sum_{n=3}^{103} n^2$$

103 ← End at $n = 103$

n^2 ← The sum of the squares of the integers

$n=3$ ← Start at $n = 3$

The symbol Σ, *sigma*, is the Greek letter, upper case S.

[1]According to historians, this story may be apocryphal and should be taken with a grain of wheat.

Sums of Finite Series

WORKED EXAMPLE 2

Write each series using sigma notation and evaluate the sum.

a) $1 + 2 + 3 + \ldots + 100$ b) $1^2 + 3^2 + 5^2 + \ldots + 31^2$

SOLUTION

a) We want the sum of the natural numbers from 1 to 100. We write: $\sum_{n=1}^{100} n$

To evaluate this sum, we could proceed as in *Worked Example* 1. Alternatively, we access sigma from the Calculus menu (F3) by pressing:

Then we enter the sequence, the variable, and the initial and final values of the variable as shown in the display.

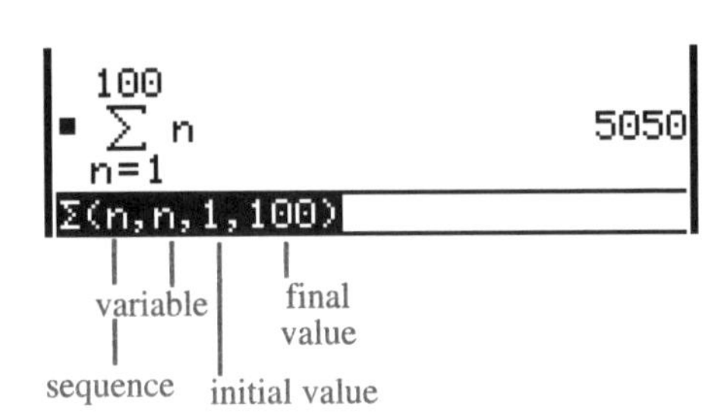

We observe in the display that the sum is 5050.

b) We write the sum as: $\sum_{n=0}^{15} (2n+1)^2$

To evaluate this sum, we can proceed as in part a) and enter the expression shown in the display on the right. Alternatively, we could use the method of *Worked Example* 1, and enter the expression as shown in the display below.

```
Σ((2n+1)^2,n,0,15)                5456
MAIN     RAD AUTO     SEQ     1/70
```

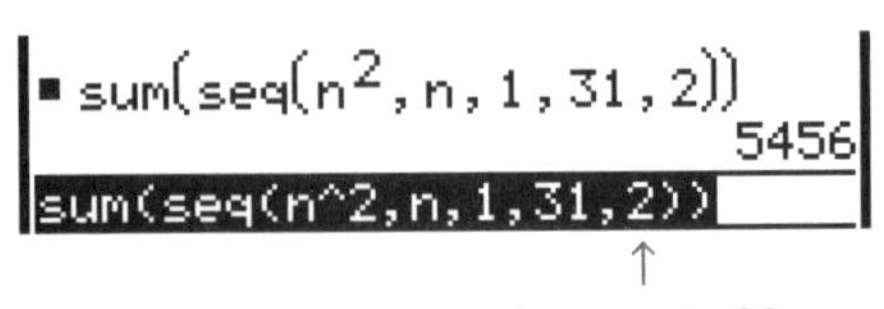

increment of 2

Sums of Infinite Series

On page 12, we saw that an infinite series can indeed have a finite sum. To evaluate infinite series, we proceed as in *Worked Examples* 1 and 2, except we use ∞ as our "final" value of *n*.

WORKED EXAMPLE 3

Write each series using sigma notation and evaluate the sum.

a) $1 + \frac{1}{2} + \frac{1}{2^2} + \ldots + \frac{1}{2^n} + \ldots$ b) $1 + \frac{1}{2^2} + \frac{1}{3^2} + \ldots + \frac{1}{n^2} + \ldots$

Can you use the **sum(seq(** commands to evaluate infinite series?

SOLUTION

a) In sigma notation, the sum is $\sum_{n=0}^{\infty} 2^{-n}$

By entering the expression shown in the command line of the display, we find that the sum of the series is 2.

```
Σ(2^-n,n,0,∞)                2
```

b) In sigma notation, the sum is $\sum_{n=1}^{\infty} n^{-2}$

By entering the expression shown in the command line of the display, we find that the sum of the series is $\pi^2/6$.

```
Σ(n^-2,n,1,∞)                π²/6
```

Exercises & Investigations

❶. Explain how Zeno's *Dichotomy* created a paradox for the ancient Greeks. What was the error in Zeno's reasoning?

❷. Explain the meaning of each of the symbols a, b, n and $u(n)$ in the expression $\sum_{n=a}^{b} u(n)$.

❸. Write the first three terms and the last term of each sum.

a) $\sum_{n=1}^{20} n^3$ b) $\sum_{n=1}^{100} 2^{-n}$ c) $\sum_{n=10}^{20} \frac{1}{n^2}$ d) $\sum_{n=1}^{20}(3n^2+2)$

❹. Write each series in sigma notation and find its sum.

a) $1+\frac{1}{2}+\frac{1}{2^2}+\ldots+\frac{1}{2^{60}}$ b) $1+\frac{1}{2}+\frac{1}{3}+\ \ldots\ +\frac{1}{100}$

c) $1+3+5+\ldots+201$ d) $2-2^3+2^5-2^7+\ldots-2^{15}$

❺. Evaluate:
(a) the sum of the squares of the integers n: $-12 \le n \le 12$.
(b) the sum of the reciprocals of the integers n, where $100 \le n \le 200$.
(c) the sum of the cubes of the reciprocals of all integers, n where $1 \le n \le 50$.

❻. An arithmetic series of n terms can be written:

$a + (a + d) + (a + 2d) + \ldots + [a + (n-1)d]$

where a is the first term and d is the common difference.

a) Express this series in sigma notation.

b) Write a formula in terms of a, d, and n for the sum of this series.

c) Use the formula in part b) to sum the multiples of 3 to 99.

❼. Find a formula for the sum of each series.

a) $\sum_{n=1}^{N} n$ b) $\sum_{n=1}^{N} n^2$ c) $\sum_{n=1}^{N} n^3$ d) $\sum_{n=0}^{N} x^n$

❽. A geometic series of n terms can be written:

$a + ar + ar^2 + ar^3 + \ldots + ar^{n-1}$

where a is the first term and r is the common ratio.

a) Express this series in sigma notation.

b) Write a formula in terms of a, r, and n for the sum of this series.

c) Use this formula to find the sum of the series:

$(1.07) + (1.07)^2 + (1.07)^3 + \ldots + (1.07)^{60}$

d) Write a formula in terms of a and r for the sum of the infinite geometric series: $a + ar + ar^2 + ar^3 + \ldots$

e) Use your formula in part d) to find the sums of these series: i) $\sum_{n=1}^{\infty}\left(\frac{1}{5}\right)^n$ ii) $\sum_{n=1}^{\infty}\left(\frac{3}{8}\right)^n$ iii) $\sum_{n=1}^{\infty} 4(1.06)^{-n}$

❾. A popular story from the history of mathematics tells of Karl Friedrich Gauss's percocity in summing a simple arithmetic series. Apparently, when he was eight years old (1785), Gauss and his schoolmates were challenged by the schoolmaster to sum the numbers from 1 to 100. Gauss astonished his teacher by scribbling the sum 5050 on his slate within minutes. It is believed that he summed the numbers by grouping them into pairs equidistant from the middle number, 50. That is, he wrote the numbers in ascending and descending order as follows:

$$\begin{array}{llll} 1\ +\ 2+\ \ 3+\ \ldots & & +48+49+50 \\ \underline{100}+\underline{99}+\ \underline{98}+ & & +\underline{53}+\underline{52}+\underline{51} \\ 101\quad 101\quad 101 & & 101\quad 101\quad 101 \end{array}$$

He reasoned that 50 sums of 101 total $50 \times 101 = 5050$.

a) Generalize Gauss's method to derive the formula for the sum of the general arithmetic series that you discovered in exercise ❻.

b) Let S denote the sum of the general geometric series of n terms. Multiply S by r and write the series corresponding to the difference between S and rS. Use this expression to deduce the formula for the sum of the general geometric series that you discovered in exercise ❽.

c) The general arithmetic-geometric series of n terms is a series of the form:

$a + (a+d)r + (a + 2d)r^2 + \ldots + [a + (n-1)d]r^{n-1}$

where the first factors of each term form an arithmetic series and the second factors form a geometric series. Use the technique in part b) to derive a formula for the sum of this series.

d) Use your TI-89 calculator to find a formula for the sum of the general arithmetico-geometric series given in part c). Use the **comDenom(** command on the Algebra menu to help simplify your formula.

Is the Harmonic Series Finite?

The sum of the reciprocals of the natural numbers is called the *harmonic series*. We write the harmonic series as: $\sum_{n=1}^{\infty}\frac{1}{n}$

↑

IS THIS SUM FINITE?

Recall from *Worked Example* 3, $\sum_{n=1}^{\infty}\frac{1}{n^2} = \frac{\pi^2}{6}$

Evaluate these sums and then conjecture whether the harmonic series has a finite sum.

a) $\sum_{n=1}^{100}\frac{1}{n}$ b) $\sum_{n=1}^{1000}\frac{1}{n}$ c) $\sum_{n=1}^{5000}\frac{1}{n}$

Give reasons to support your conjecture.

Exploration 3 Areas as Limits of Series: Exhaustion

CORBIS-BETTMANN

Give me a place to stand on and I will move the earth.
Archimedes 287 – 212 B.C.

Archimedes (287-212 B.C.) is described by some historians as the mathematician who anticipated calculus almost two millenia before its formal development by Newton and Leibniz. When Isaac Newton made his famous statement, *If I have seen further than others, it is because I stood on the shoulders of giants,* he was probably referring to Archimedes among others.

From the time of the Pythagoreans (c. 550 B.C.) the ancient Greeks struggled with the concept of the infinite. They had discovered the irrationality of $\sqrt{2}$, but they regarded only the rationals as "real" numbers. The Zeno paradoxes (c. 430 B.C.) further confounded their investigation of the infinite. It was Archimedes who extended the so-called *method of exhaustion* (described below) and moved mathematics a giant step toward the concept of the limit.

The main elements in the concept of the limit are seen in Archimedes' method for estimating the numerical value of π. It was known that the circumference of a circle of radius 1 is a constant, denoted 2π, but the numerical value of π was not known to any degree of accuracy. Archimedes' method for estimating the value of π was as follows:

❶. **Observation:** The perimeter $q(n)$ of a polygon of n sides circumscribed about a circle exceeds the circumference of the circle.

❷. **Observation:** The perimeter $p(n)$ of a polygon of n sides inscribed in a circle is less than the circumference of the circle.

❸. **Observation:** As n increases, the sequence $q(n)$ decreases to the circumference of the circle and the sequence $p(n)$ increases to the circumference of the circle.

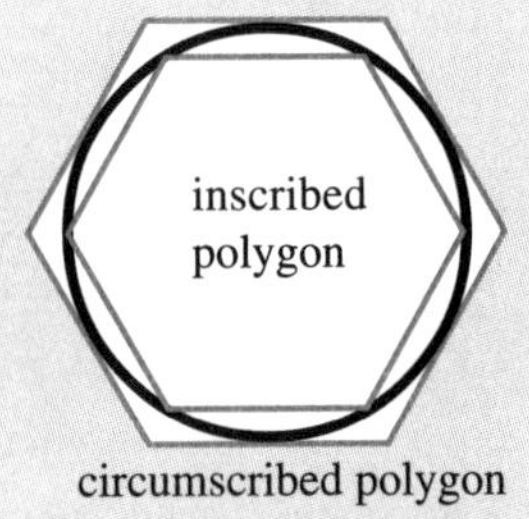

Archimedes' Idea

By calculating the perimeters of the inscribed and circumscribed polygons of n sides (for large n), we can find upper and lower bounds which closely approximate the circumference of the unit circle. This gives us an approximation of 2π, and hence π. Using $n = 96$, Archimedes was able to obtain the following upper and lower bounds for π.

$$3\frac{10}{71} < \pi < 3\frac{1}{7}$$

We show in *Worked Example* 1 that the expressions $p(n)$ and $q(n)$ for $n = 96$ are respectively, $2(96)\sin(\pi/96)$ and $2(96)\tan(\pi/96)$. That is, the estimates of π we obtain from the 96-sided polygon are respectively $96\sin(\pi/96)$ and $96\tan(\pi/96)$. When we compute these values on our TI-89, we obtain to three decimal places :

$$3.141 < \pi < 3.143$$

This is what we get when we express Archimedes' result to 3 decimal places.

Archimedes' achievement is even more remarkable when we realize that trigonometry had not yet been invented and decimal notation was not available! He began with a regular hexagon, and used the simple properties of the 30-60-90 triangle to calculate $p(6)$ and $q(6)$. He used proportionality relationships in right triangles to deduce $p(12)$ and $q(12)$ and so on for $n = 24$, 48 and 96. This is equivalent to using the modern half-angle trig formulas. His ability to execute these complex computations is a tribute to his tenacity and his genius.

WORKED EXAMPLE 1

Write expressions for the perimeters $p(n)$ and $q(n)$ of the inscribed and circumscribed n-sided regular polygons and find the limit of each expression as n approaches infinity.

SOLUTION

The top diagram shows one of n triangles into which an n-sided regular inscribed polygon has been divided. The angle at the center of the circle is $1/n$ of a complete rotation 2π; i.e., $2\pi/n$. Therefore, the angle in the right triangle is π/n, so the opposite side has length $\sin(\pi/n)$. That is, the length of AB is $2\sin(\pi/n)$. The perimeter of the inscribed polygon of n sides is therefore,

$$p(n) = 2n\sin(\pi/n)$$

The lower diagram shows one of the n triangles into which an n-sided regular circumscribed polygon has been divided. The length CD is $2\tan(\pi/n)$. The perimeter of the circumscribed polygon of n sides is therefore,

$$q(n) = 2n\tan(\pi/n)$$

The lower and upper bounds for the circumference of the unit circle are therefore, $\lim_{n\to\infty} p(n)$ and $\lim_{n\to\infty} q(n)$ respectively.

To evaluate $\lim_{n\to\infty} p(n)$, we press:      2nd

[sin] 2nd [π] ÷ n) , n ,  [∞]  ENTER

We obtain the display below. We observe that $\lim_{n\to\infty} p(n) = 2\pi$.

In exercise ⑨ on page 24, you will discover how to evaluate this limit without your TI-89.

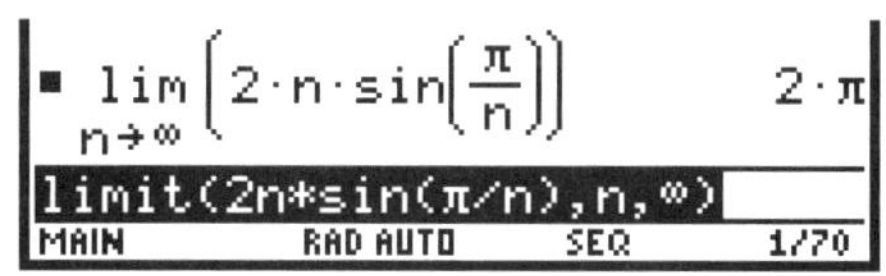

Similarly, we calculate the limit of q(n) as $n \to \infty$ and obtain $\lim_{n\to\infty} q(n) = 2\pi$.

In exercise ⑨ on page 24, you will discover how to evaluate this limit without your TI-89.

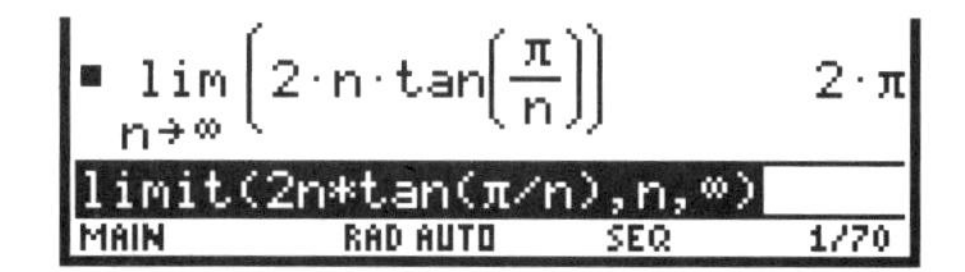

Since $p(n)$ < circumference < $q(n)$ for all n, then,

$$\lim_{n\to\infty} p(n) \le \text{circumference} \le \lim_{n\to\infty} q(n).$$

Since both these limits are 2π, then the circumference is 2π.

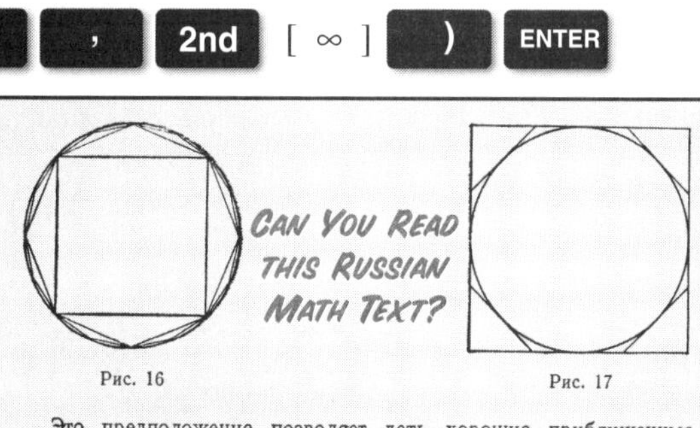

Рис. 16 Рис. 17

Это предположение позволяет дать хорошие приближенные оценки числа π. Еще Архимед установил, что периметр P_{96} вписанного в окружность диаметра d правильного девяностошестиугольника удовлетворяет неравенству

$$P_{96} > 3\frac{10}{71}d,$$

а периметр Q_{96} описанного правильного девяностошестиугольника — неравенству

$$Q_{96} < 3\frac{1}{7}d.$$

Отсюда вытекает (если принять высказанное выше допущение), что

$$3\frac{10}{71} < \pi < 3\frac{1}{7}.$$

Из соображений подобия вытекает, что

$$P_n = p_n d, \; Q_n = q_n d, \qquad (2)$$

где p_n и q_n — периметры соответственно вписанных и описанных правильных многоугольников для окружности диаметра $d = 1$.
Оказывается, что последовательности

$$p_3, p_4, p_5, \ldots, p_n, \ldots,$$
$$q_3, q_4, q_5, \ldots, q_n, \ldots$$

имеют общий предел. Этот предел и есть число π:

$$\pi = \lim_{n\to\infty} p_n = \lim_{n\to\infty} q_n. \qquad (3)$$

WORKED EXAMPLES

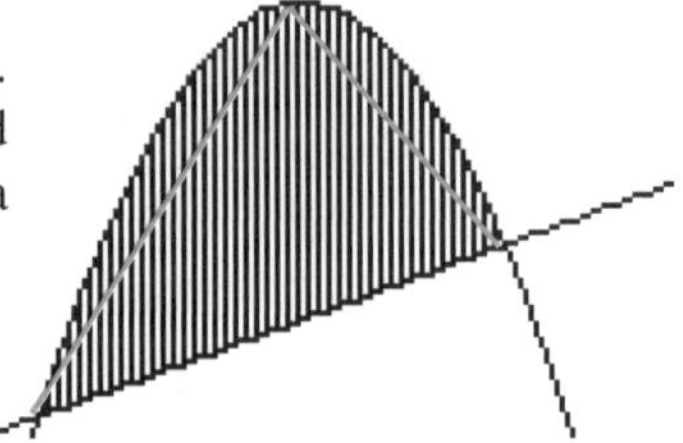

Archimedes moved to the threshold of integral calculus in his use of the *method of exhaustion*. This involves approximating areas by partitioning them into n rectangles or triangles and taking the limit as n becomes large. Using this technique, Archimedes proved the area of a parabolic segment is 4/3 the area of a triangle with the same base and height.

WORKED EXAMPLE 2

Find the area of the first quadrant enclosed by the parabola with equation $y = -x^2 + 4$:
a) by the method of exhaustion.
b) by evaluating a numerical integral.

SOLUTION

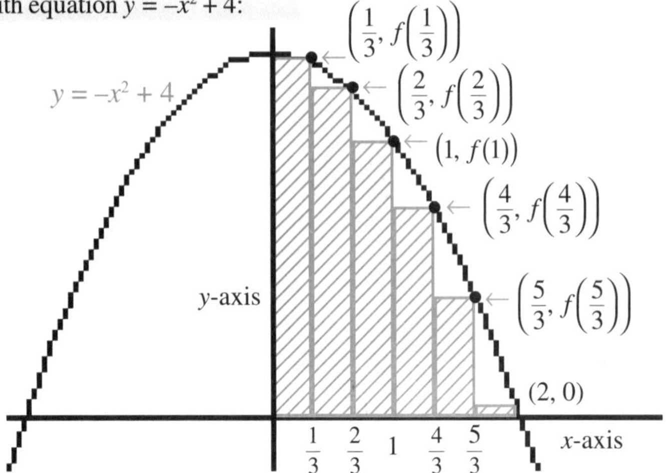

The total area of the shaded rectangles is denoted S(6). It is a lower bound estimate of the area under the parabola

a) To estimate the area under the parabola and in the interval $0 \le x \le 2$, we partition the interval into 6 parts of width 1/3. We then construct rectangles which stretch from the x-axis to the curve. The area S(6) of these 6 rectangles is a lower bound estimate of the area under the parabola. We write:

$$S(6) = \frac{1}{3}\left[f\left(\frac{1}{3}\right) + f\left(\frac{2}{3}\right) + f(1) + f\left(\frac{4}{3}\right) + f\left(\frac{5}{3}\right) + f(2)\right]$$

$$= \frac{1}{3}\sum_{k=1}^{6} f\left(\frac{k}{3}\right) = \frac{1}{3}\sum_{k=1}^{6}\left[-\left(\frac{k}{3}\right)^2 + 4\right] \text{ or } \frac{125}{27} \approx 4.69\ldots$$

We see that S(6) gives us an estimate of 4.69 for the area under the parabola and in the first quadrant. The error between this and the true area is the combined area of all the small white triangles under the curve. The method of exhaustion involves eliminating this error by taking finer and finer partitions. That is, we calculate S(n) and then take the limit of S(n) as $n \to \infty$.

In general, the sum of the areas of n rectangles which partition the region $0 \le x \le 2$ is S(n):

$$S(n) = \frac{2}{n}\left[f\left(\frac{2}{n}\right) + f\left(\frac{4}{n}\right) + \ldots + f\left(\frac{2k}{n}\right) + \ldots + f\left(\frac{2nk}{n}\right)\right] = \frac{2}{n}\sum_{k=1}^{n} f\left(\frac{2k}{n}\right) = \frac{2}{n}\sum_{k=1}^{n}\left[-\left(\frac{2k}{n}\right)^2 + 4\right]$$

To evaluate S(n), we enter the expression shown in the command line of the display. The display yields:

$$S(n) = \frac{4(4n^2 - 3n - 1)}{3n^2}$$

```
■ 2/n · Σ(k=1..n) [ -(2·k/n)^2 + 4 ]
                         4·(4·n^2 - 3·n - 1)
                         -------------------
                               3·n^2
2/n*Σ(-(2k/n)^2+4,k,1,n)
```

The lower bound for the area is $\lim_{n\to\infty} S(n) = \lim_{n\to\infty} \frac{4(4n^2 - 3n - 1)}{3n^2}$.

To evaluate this limit, we proceed as in *Worked Example* 1 to obtain the value 16/3 shown in the display. Similarly, we find an upper bound for the area is $\lim_{n\to\infty} T(n) = \frac{16}{3}$ where $T(n)$ is the area of rectangles which enclose the parabola. These bounds both approach 16/3, so the area must approach 16/3.

```
■ lim(n→∞) [ 2/n · Σ(k=1..n) [ -(2·k/n)^2 + 4 ] ]
                                           16/3
...(-(2k/n)^2+4,k,1,n),n,∞)
```

b) Your TI-89 calculator has a *numerical integral* function that performs the method of exhaustion automatically. To access this function, press: **F3** **B** and enter the commands in the display to obtain area 5.33…as in part a).

```
■ nInt(-x^2 + 4, x, 0, 2)
                        5.33333
nInt(-x^2+4,x,0,2)
```

EXERCISES & INVESTIGATIONS

1. a) Write a few sentences to explain the method of exhaustion to a friend who missed class.

b) If $S(n)$ and $T(n)$ are respectively lower and upper bounds to the true area under a curve, such that

$$S(n) \le \text{True Area} \le T(n), \text{ for all } n$$

and $\lim_{n\to\infty} S(n) = \lim_{n\to\infty} T(n)$

Is it true that

$$\text{True Area} = \lim_{n\to\infty} S(n) = \lim_{n\to\infty} T(n)\ ?$$

Explain why or why not.

2. The diagrams below show respectively one of the n triangles into which the inscribed and circumscribed n-sided polygons of a unit circle are divided. (See *Example* 1.)

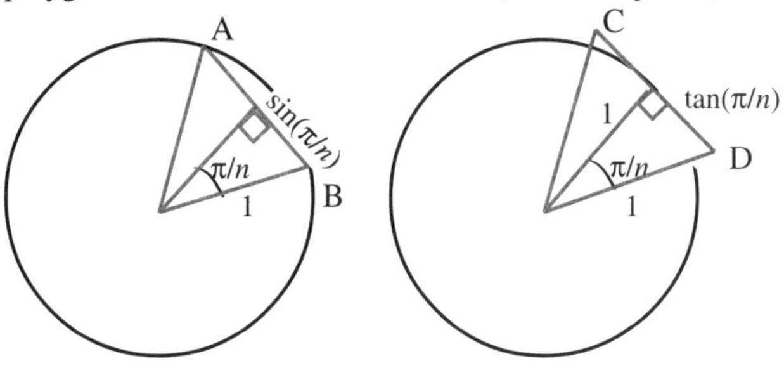

a) Write an expression (as a function of n) for P(n) the area of the inscribed n-sided polygon of the unit circle.

b) Write an expression for Q(n) the area of the circumscribed n-sided polygon of the unit circle.

c) Find the limit of each expression as n approaches ∞.

d) Interpret the meaning of your result in part c).

3. The diagram shows a rectangular partition which provides an upper bound for the area in the first quadrant enclosed by parabola defined by $y = -x^2 + 4$.

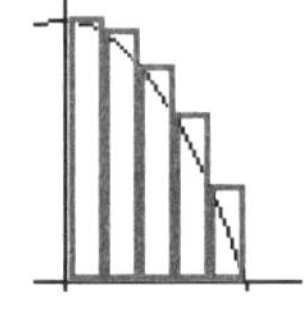

a) Write a series for T(5) the sum of the areas of the rectangles forming the upper bound to the area.

b) Use your TI-89 to sum the series you obtained in part a).

c) Write a general expression for T(n)

d) Evaluate $\lim_{n\to\infty} T(n)$.

4. Use the numerical integration command to approximate the area:

a) below the x-axis and above the parabola $y = x^2 - 4$.

b) below the curve $y = \log_{10} x$ for $1 \le x \le 3$.

c) below the curve $y = 1/(x + 5)$ for $0 \le x \le 4$.

d) below the curve $y = \left(1+\sqrt[3]{x}\right)$ for $1 \le x \le 8$.

5. The equation of an ellipse with semi-major axis a and semi-minor axis b is $b^2x^2 + a^2y^2 = a^2b^2$. Follow the steps below to find a formula in terms of a and b for the area inside this ellipse.

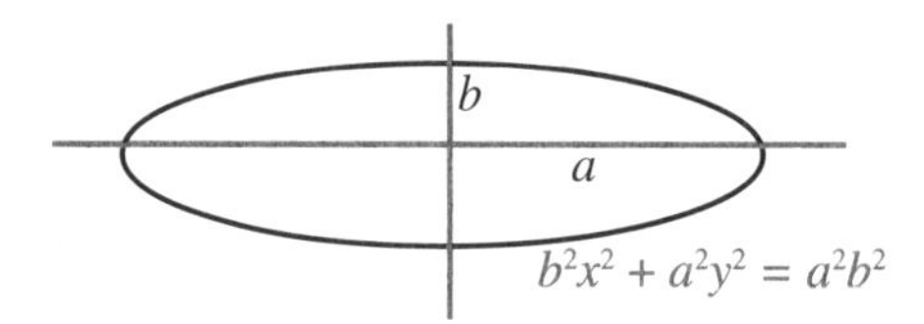

a) Write an expression in terms of a and b for S(n), the lower rectangular approximation to the area in the first quadrant enclosed by the ellipse. Write S(n) in the form:

$$S(n) = abC(n)$$

Observe that C(n) is the area of the unit circle in the first quadrant. In exercise **2**, you proved that the area of the unit circle is π, and so C(n) = $\pi/4$. Therefore, S(n) = $\pi ab/4$. Hence, the area inside this ellipse is πab. However, we shall proceed without using the result obtained in exercise **2**.

b) Can you simplify your expression for C(n) using your TI-89? Explain why or why not.

c) Can you evaluate the limit of C(n) as $n \to \infty$? Explain why or why not.

d) To approximate the value of C(n), we construct our rectangular approximations of the unit circle in the first quadrant .

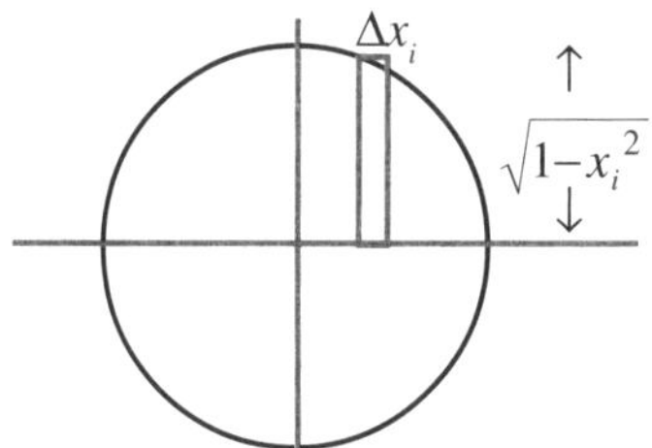

We then perform what is called a *numerical integration* using the command **nInt(** on the calculus menu (F3). To do this, we write the sum C(n) in terms of the variable x.

$$\lim_{n\to\infty} \sum_{i=1}^{n} \sqrt{1-x_i^2}\ \Delta x_i$$

where Δx_i is the width of the ith rectangle and $\sqrt{1-x_i^2}$ its height. When the **nInt(** command is applied, your TI-89 computes this sum for a sample of values x_i and continues to sample until the sum appears to approach a limit. It then yields a numerical approximation to the true area. Perform the numerical integration of $\sqrt{1-x^2}$ for x between 0 and 1 by entering on the command line the commands shown in this display. (Observe this yields $\pi/4$ as indicated above.)

```
nInt(√(1-x^2),x,0,1)
```

Therefore, the area inside the ellipse defined by $b^2x^2 + a^2y^2 = a^2b^2$ is given by πab.

LIMITS OF INFINITE GEOMETRIC SERIES

In *Exploration* 2, we used the **limit(** command to show that

$$\lim_{n\to\infty}\left(\frac{1}{2}+\frac{1}{4}+\frac{1}{8}+\ldots+\frac{1}{2^n}\right)=1$$

How could we have evaluated this limit without using technology?

In part b) of investigation ❾ on page 15, you may have discovered that the sum of the geometric series,

$$a + ar + ar^2 + ar^3 + \ldots + ar^{n-1},$$

where a is the first term and r is the common ratio, is

$$S(n)=\frac{a(r^n-1)}{r-1} \quad (*)$$

To evaluate the limit of $S(n)$ as $n\to\infty$, write $S(n)$ as:

$$S(n)=\frac{ar^n}{r-1}-\frac{a}{r-1}$$

When $|r|<1$, the term r^n approaches 0 for large values of n.

That is, $\lim_{n\to\infty} r^n=0$. Hence, the numerator of the first term of $S(n)$ decreases to 0, and we have,

$$\lim_{n\to\infty} S(n)=\frac{a}{1-r} \quad \text{for } |r|<1$$

We write:

$$\boxed{\lim_{n\to\infty}\left(a+ar+ar^2+\ \ldots\ +ar^{n-1}\right) = \frac{a}{1-r} \quad \text{where } |r|<1}$$

← **LIMIT OF AN INFINITE GEOMETRIC SERIES**

"UNDERSTANDING CREATION AND INFINITY IS WITHIN OUR GRASP... JUST AS SOON AS EVOLUTION PROVIDES US WITH LARGER BRAINS."

In words, the sum corresponding to an infinite geometric series with common ratio $|r|<1$, is the first term divided by $1-r$. Therefore, the sum for the series above is given by

$$\lim_{n\to\infty}\left(\frac{1}{2}+\frac{1}{4}+\frac{1}{8}+\ldots+\frac{1}{2^n}\right)=\frac{\frac{1}{2}}{1-\frac{1}{2}} \text{ or } 1.$$

To obtain the nth partial sum $S(n)$ for this series we substitute $a = 1/2$ and $r = 1/2$ into equation $(*)$ above. This yields $S(n) = 1 - 2^{-n}$. The graph of $S(n)$ for $0 \le n \le 20$ is shown in the display.

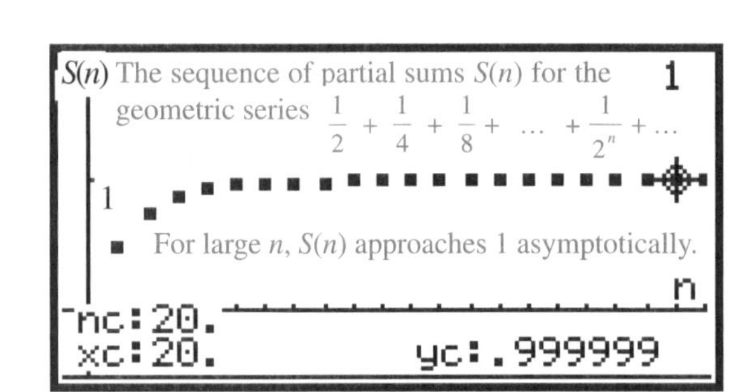

In *Worked Example* 2 on page 18, we applied the method of exhaustion to find the area enclosed by the parabola with equation $y = -x^2 + 4$. This required that we evaluate

$$\lim_{n\to\infty} S(n) = \lim_{n\to\infty}\frac{4(4n^2-3n-1)}{3n^2}$$

Using the **TI-89 limit(** command, we found that this limit is 16/3. Alternatively, we could have divided the numerator and denominator of this rational expression by n^2 to obtain

$$\lim_{n\to\infty} S(n) = \lim_{n\to\infty}\frac{4\left(4-\frac{3}{n}-\frac{1}{n^2}\right)}{3}$$

For large values of n the terms $-\frac{3}{n}$ and $-\frac{1}{n^2}$ approach arbitrarily close to 0, so $\lim_{n\to\infty} S(n)=\frac{16}{3}$.

WORKED EXAMPLES

When evaluating the limit of an infinite series, we are actually finding the limit of the infinite sequence of partial sums S(n), where S(n) is the sum of the first n terms. Since a sequence is merely a function defined on the positive integers, we can extend the techniques for evaluating limits of series to the evaluation of limits of functions.

> **Definition:** We say that the limit of a function $f(x)$ as x approaches c is L (written $\lim_{x \to c} f(x) = L$), if $f(x)$ approaches arbitrarily close to the value of L as x approaches c.

In the worked examples, we use the principles, $\lim_{x\to\infty} x^m = \infty$ and $\lim_{x\to\infty} \frac{1}{x^m} = 0$ for $m > 0$ (✻)
to evaluate the limits of polynomials and of rational functions.

WORKED EXAMPLE 1

Calculate each limit with and without your TI-89.

a) $\lim_{x\to\infty}(3x^2 - 7x + 4)$ b) $\lim_{x\to -2}(4x^3 - 9x^2 + 4)$ c) $\lim_{x\to -3} \frac{5x^2+7x+3}{x^3+2x^2-9x+11}$

SOLUTION

To evaluate these limits, we will assume some properties of limits which you will prove in the exercises on page 24, namely;

- the limit of a sum of functions is the sum of the limits,
- the limit of a product of two functions is the product of the limits
- the limit of a quotient of two functions is the quotient of the limits (provided the limit in the denominator is not zero.)

a) As x increases without limit $3x^2 \to \infty$ and $-7x \to -\infty$, so the sum appears to approach $\infty - \infty$ which is indeterminate. That is, direct substitution of ∞ for x leads to a meaningless expression. To evaluate this limit, divide $3x^2 - 7x + 4$ by the highest power of x as follows

$$\lim_{x\to\infty}(3x^2 - 7x + 4) = \lim_{x\to\infty} x^2 \times \lim_{x\to\infty}\left(3 - \frac{7}{x} + \frac{4}{x^2}\right)$$

Applying the principles in (✻), we find $\lim_{x\to\infty} x^2 = \infty$ and $\lim_{x\to\infty}\left(3 - \frac{7}{x} + \frac{4}{x^2}\right) = 3$

Since the limit of a product is the product of the limits,

$\lim_{x\to\infty}(3x^2 - 7x + 4) = \infty$. The display verifies this result.

```
■ lim (3·x^2 - 7·x + 4)        ∞
  x→∞
limit(3x^2-7x+4,x,∞)
```

b) Substituting – 2 for x yields, $f(-2) = -64$ so,

$$\lim_{x\to -2}(4x^3 - 9x^2 + 4) = -64$$

```
■ lim (4·x^3 - 9·x^2 + 4)
  x→ -2
                              -64
limit(4x^3-9x^2+4,x,-2)
```

c) Substituting –3 for x yields, $f(-3) = 27/29$ so,

$$\lim_{x\to -3} \frac{5x^2+7x+3}{x^3+2x^2-9x+11} = \frac{27}{29}$$

```
■ lim   (    5·x^2 + 7·x + 3     )
  x→ -3 ( x^3 + 2·x^2 - 9·x + 11 )
                            27/29
…)/(x^3+2x^2-9x+11),x,-3)
```

Worked Example 1 suggests that the evaluation of $\lim_{x \to c} f(x)$ merely involves evaluating $f(c)$. The next example shows that sometimes $f(c)$ yields an indeterminate form such as $\frac{0}{0}$ or $\frac{\infty}{\infty}$. In such cases, we can often remove the indeterminacy by dividing the numerator and denominator by a common element.

WORKED EXAMPLE 2

Calculate each limit with and without your TI-89 Graph the corresponding function of x.

a) $\lim_{x \to \infty} \frac{5x^2+7x+3}{x^3+2x^2-9x+11}$ b) $\lim_{x \to 3} \frac{x^3-10x^2+31x-30}{x-3}$ c) $\lim_{x \to -\infty} \frac{4x^3+9x-6}{3x^3-x^2-7x+5}$

SOLUTION

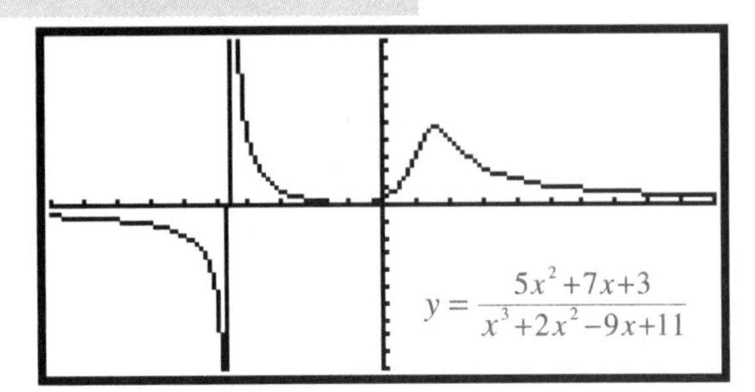

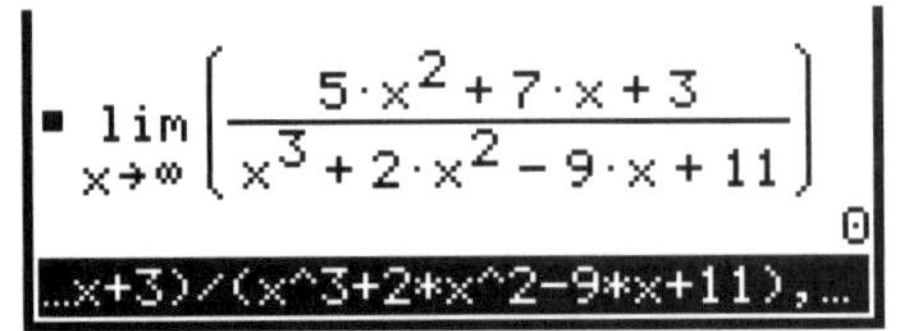

a) Substituting ∞ for x, yields the indeterminate form $\frac{\infty}{\infty}$.

Proceeding as in part a) of *Worked Example* 1, we divide numerator and denominator by the highest power of x i.e. x^3.

$$\lim_{x \to \infty} \frac{5x^2+7x+3}{x^3+2x^2-9x+11} = \lim_{x \to \infty} \frac{\frac{5}{x}+\frac{7}{x^2}+\frac{3}{x^3}}{1+\frac{2}{x}-\frac{9}{x^2}+\frac{11}{x^3}} = 0$$

We observe that as $x \to \infty$, the numerator approaches 0 and the denominator approaches 1 so the quotient approaches 0. The graph of

$$y = \frac{5x^2+7x+3}{x^3+2x^2-9x+11}$$

(shown in the display) reveals that this function has a singularity at $x \approx -4.5$ and a local maximum at $x \approx 1.6$ and that it does, in fact, approach the value 0 as x increases without limit. The **limit (** command also verifies the computation above.

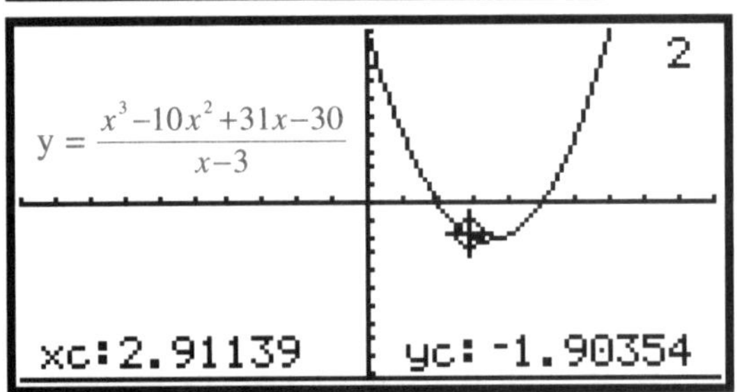

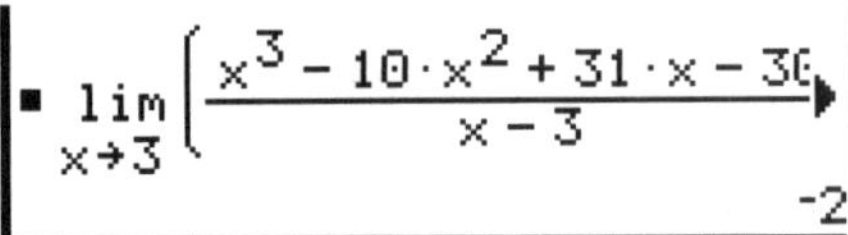

b) Substituting 3 for x, yields the indeterminate form $\frac{0}{0}$. This indicates that there is at least one factor $(x-3)$ in both the numerator and denominator. We therefore, factor both and cancel common factors. Then we take the limit.

$$\lim_{x \to 3} \frac{x^3-10x^2+31x-30}{x-3} = \lim_{x \to 3} \frac{(x-3)(x-5)(x-2)}{x-3} = -2$$

The graph of

$$y = \frac{x^3-10x^2+31x-30}{x-3}$$

reveals that as x gets very close to 3, y approaches -2.

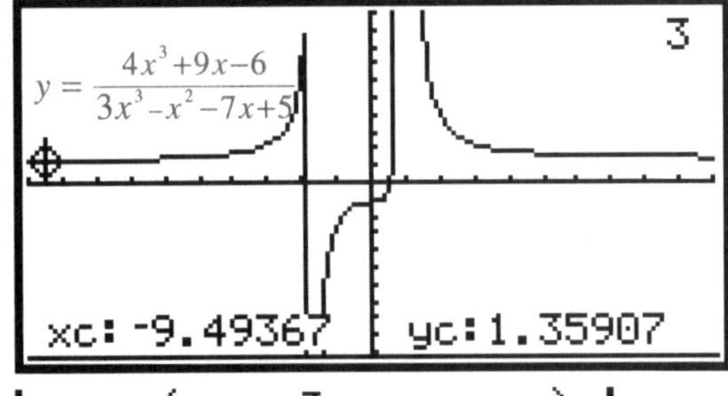

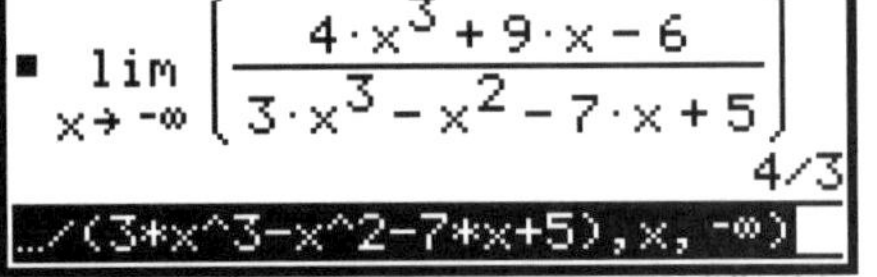

c) Substituting $-\infty$ for x yields the indeterminate form $\frac{-\infty}{-\infty}$. Division of numerator and denominator by x^3 and taking the limits as $x \to \infty$ yields

$$\lim_{x \to -\infty} \frac{4x^3+9x-6}{3x^3-x^2-7x+5} = \lim_{x \to -\infty} \frac{4+\frac{9}{x^2}-\frac{6}{x^3}}{3-\frac{1}{x}-\frac{7}{x^2}+\frac{5}{x^3}} = \frac{4}{3}$$

The graph of $y = \frac{4x^3+9x-6}{3x^3-x^2-7x+5}$ suggests the limit $\approx 4/3$.

WORKED EXAMPLES

When we have quotients involving functions which are not polynomials, we cannot remove indeterminacy by factoring.

WORKED EXAMPLE 3

Evaluate $\lim\limits_{x\to 0}\dfrac{\sin x}{x}$

To put the TI-89 in radian mode press:

MODE ▼ ▼ ▼ ▶ 1 ENTER

SOLUTION

To estimate this limit, we graph the function $y=\dfrac{\sin x}{x}$, with the window variables as shown and the calculator in radian mode. We then trace along the graph as close as possible to the point with $x = 0$, as shown in the display. The display suggests that the limit is 1 because the y-coordinates seem to approach arbitrarily closely to 1 as the x-coordinate approaches 0. However, this does not *prove* the limit is 1, because the graph might have a singularity at $x = 0$.

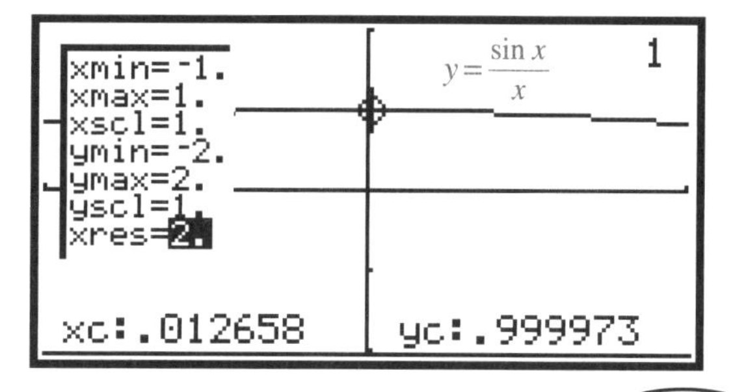

To deduce the value of the limit, we construct the triangles which inscribe and circumscribe the sector of a unit circle with sector angle $2x$. We observe:

Area of circumscribed triangle: $\tan x$
Area of inscribed triangle: $\sin x \cos x$
Area of the sector $\dfrac{2x}{2\pi}\times\pi$ *or* x

1
2sin x
2tan x
x
x
1

Area of inscribed triangle < area of sector < area of circumscribed triangle.

That is: $\sin x\cos x < x < \tan x$

Division by $\sin x$ yields: $\cos x < \dfrac{x}{\sin x} < \dfrac{1}{\cos x}$

Taking reciprocals, we obtain $\cos x < \dfrac{\sin x}{x} < \dfrac{1}{\cos x}$.

Therefore, $\lim\limits_{x\to 0}\cos x \le \lim\limits_{x\to 0}\dfrac{\sin x}{x} \le 1$.

Since $\lim\limits_{x\to 0}\cos x = 1$, then $\lim\limits_{x\to 0}\dfrac{\sin x}{x} = 1$.

WORKED EXAMPLE 4

The world population (in billions), since 1900 A.D. can be modeled by the logistic function $P(t)$, where t denotes the number of years since 1900 and $P(t)$ is given by

$$P(t)=\frac{78.12}{6.3+102e^{-0.02817t}}$$

Using this model, calculate the population in 1950 and predict the world population in 2000, 2050, and 2100. Calculate the limiting population, $\lim\limits_{t\to\infty}P(t)$.

SOLUTION

We enter the function $P(t)$ as $y1(x)$ in the TI-89. Then on the home screen, we evaluate $y1(50)$, $y1(100)$, $y1(150)$, and $y1(200)$, to obtain the respective populations, 2.5, 6.3 and 10.0 billions. To evaluate $\lim\limits_{t\to\infty}P(t)$, we observe that $\lim\limits_{t\to\infty}e^{-0.02817t}=0$, so $\lim\limits_{t\to\infty}P(t)=\dfrac{78.12}{6.3}=12.4$.
The limiting population is 12.4 billion.

Is the Population Explosion Out of Control?

In 1845 Belgian scientist P. F. Verhulst modeled the world population P as a function of the time t using the so-called *logistic function* $P(t)$. This graph of $P(t)$ shows that the population approaches closer and closer to the limiting value $L = 12.4$ as t increases, but $P(t)$ never reaches this value while t is finite. We say that $P(t)$ approaches the limit L *asymptotically* as t approaches infinity.

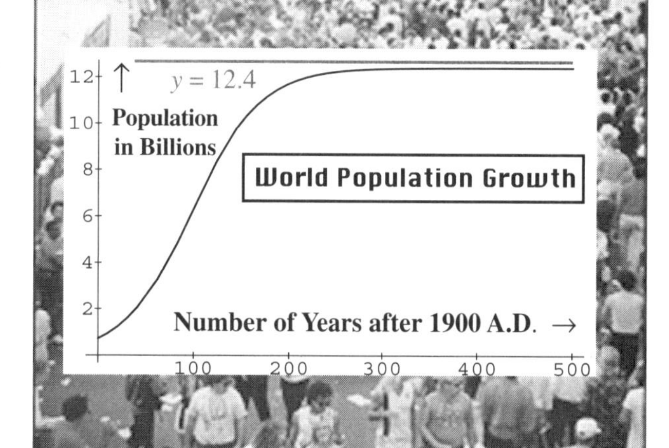

Exercises & Investigations

1. a) What is meant by the expression "n^{th} partial sum" for a series?

b) How is the nth partial sum of an infinite series related to the sum for the series?

2. a) If $f(x)$ is a function of x and c is some real number, what is meant by $\lim_{x \to c} f(x) = L$?

b) What does it mean to say that the value of $f(x)$ is "indeterminate" at a particular point?

c) What is the cause of the indeterminate form $\frac{0}{0}$ in a rational function, and how can it be removed in the calculation of a limit?

3. Given $f(x) = a_n x^n + a_{n-1}x^{n-1} + \ldots + a_1 x + a_0$ and $g(x) = b_m x^m + b_{m-1}x^{m-1} + \ldots + b_1 x + b_0$,

evaluate $\lim_{x \to \infty} \frac{f(x)}{g(x)}$ for each of the following cases.

a) $n > m$ b) $n < m$ c) $n = m$

4. a) Evaluate the partial sum $S(n)$ for the geometric series with first term a and common ratio r for these values of a, r, and n.

i) $a = -4$, $r = 3$, and $n = 9$
ii) $a = 7$, $r = 1/4$, and $n = 15$
iii) $a = 5$, $r = 2/3$, and $n = 30$

b) Graph the series of partial sums from $S(1)$ to $S(n)$ for each of the geometric series in part a).

5. a) Let $S(n)$ denote the nth partial sum of an infinite geometric series and let a and r denote the first term and common ratio respectively. Evaluate $\lim_{n \to \infty} S(n)$ for these values of a and r: i) $a = 6$, $r = 7/8$ ii) $a = 4$, $r = -3/4$

b) Graph each series of partial sums $S(n)$ for $n = 1, 2, \ldots 20$ and trace along your graph to obtain $S(20)$.

6. Evaluate each limit. Use your TI-89 to check.

a) $\lim_{x \to \infty} (5x^2 - 2x + 17)$ b) $\lim_{x \to 3} (3x^3 - 4x^2 + 9)$

c) $\lim_{x \to -\infty} \frac{7x^3 + 2x^2 - 3x + 13}{2x^3 - 7x^2 + 5x + 11}$ d) $\lim_{x \to -5} \frac{x^3 + 5x^2 - 2x - 10}{x^3 + 11x^2 + 23x - 35}$

7. Evaluate each limit. Use your TI-89 to check.

a) $\lim_{x \to -1} \frac{7x^3 + 14x^2 + 7x}{x^2 + 2x + 1}$ b) $\lim_{x \to \sqrt{2}} \frac{x^3 - 3x^2 - 2x + 6}{x^4 - 4x^3 + 2x^2 + 8x - 8}$

c) $\lim_{x \to 0} \frac{\sin x}{3x}$ d) $\lim_{x \to 0} \frac{\sin 3x}{x}$

8. Evaluate $\lim_{x \to \infty} x^2[1 - \cos(x^{-1})]$ and $\lim_{x \to 0} x^2[1 - \cos(x^{-1})]$.

9. a) Use $\lim_{x \to 0} \frac{\sin x}{x} = 1$ to evaluate $\lim_{x \to 0} \frac{\sin kx}{x} = 1$ where k is a real number.

b) Use the result in part a) to deduce the limits displayed on the TI-89 in *Worked Example* 1 p. 17, that is,

$$\lim_{n \to \infty} 2n \sin\left(\frac{\pi}{n}\right) \text{ and } \lim_{n \to \infty} 2n \tan\left(\frac{\pi}{n}\right)$$

10. Graph the function defined by $y = \frac{(6x+1)^4 - (6x)^4}{x^4}$.

a) Trace along your graph to evaluate:

$$\lim_{x \to 0} \frac{(6x+1)^4 - (6x)^4}{x^4}$$

Does this limit exist? Explain why or why not.

b) Trace along your graph to estimate the value of:

$$\lim_{x \to \infty} \frac{(6x+1)^4 - (6x)^4}{x^4}$$

c) Use the **expand(** command from the Algebra menu (F2) to expand $\frac{(6x+1)^4 - (6x)^4}{x^4}$.

d) Use your result in c) to calculate the exact value of $\lim_{x \to \infty} \frac{(6x+1)^4 - (6x)^4}{x^4}$ and compare with your estimate.

11. Evaluate $\lim_{x \to 0} \frac{\sqrt{30x+30} - \sqrt{30}}{x}$.

12. a) Use the **MODE** key to select SEQUENCE mode.

Then graph $u1(n) = \left(1 + \frac{1}{n}\right)^n$ for $1 \le n \le 20$, in the window $-1 \le x \le 21$; $-0.5 \le y \le 3$. Evaluate $u1(60)$.

b) Evaluate $\lim_{n \to \infty} u1(n) = \left(1 + \frac{1}{n}\right)^n$. Express this limit as a decimal number. This number, denoted by e is the base of the natural logarithms. (To convert e to decimal form, press: **◆** **ENTER**.)

c) Use the **MODE** key to select FUNCTION mode.

Graph $y1(x) = \left(1 + \frac{1}{x}\right)^x$ for $-1 \le x \le 20$, $-1 \le y \le 3$.

Evaluate $\lim_{x \to \infty} y1(x) = \left(1 + \frac{1}{x}\right)^x$.

d) Graph $y2(x) = (1 + x)^{\frac{1}{x}}$ for $-1 \le x \le 10$, $-1 \le y \le 5$. Trace along the graph to obtain an estimate of

$$\lim_{x \to 0} (1 + x)^{\frac{1}{x}}.$$

e) Use your results in parts c) and d) to express these limits as functions of e.

(i) $\lim_{x \to 0} (1 + ax)^{\frac{b}{x}}$ (ii) $\lim_{x \to \infty} \left(1 + \frac{a}{x}\right)^{bx}$

The Concept of The Derivative

Mathematical Concepts

- The derivative as a limit
- instantaneous velocity & acceleration
- binomial theorem for fractional exponents
- the tangent as a limit of secants
- the derivative as the slope of a tangent
- The Power Rule for derivatives
- evaluating derivatives of polynomials
- tangent lines
- extrema of functions
- optimization problems
- trajectories of projectiles
- distance-time graphs

TI-89 Commands

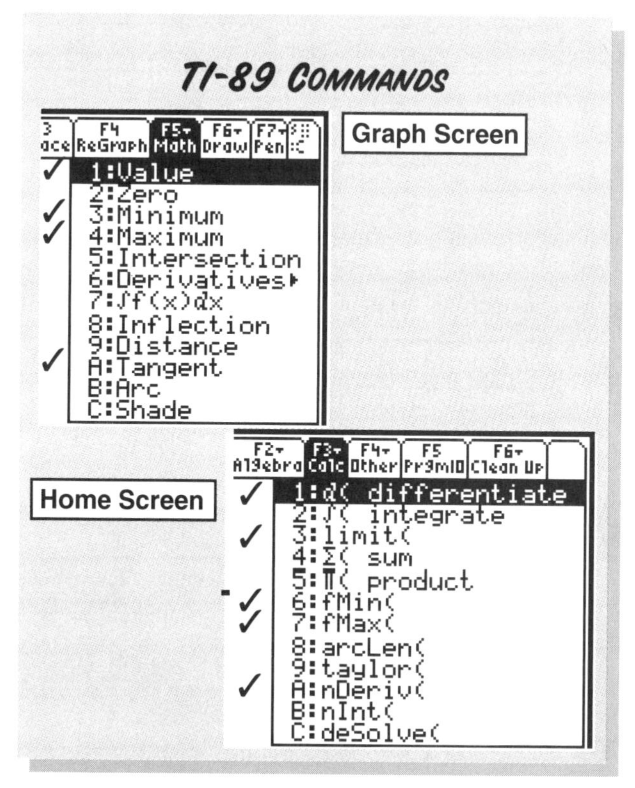

DISTANCE, RATE & TIME

CORBIS-BETTMANN

Galileo Galilei 1564 – 1642

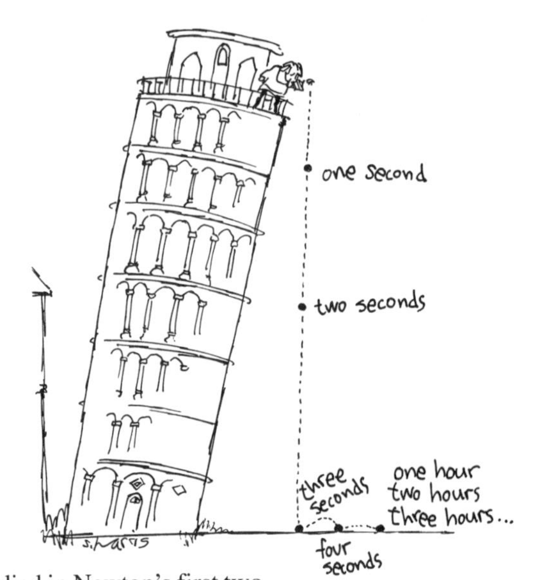

We stated on page 16, "When Isaac Newton made his famous statement, *If I have seen further than others, it is because I have stood on the shoulders of giants*, he was probably referring to Archimedes among others." Certainly, one of the "others" he had in mind was Galileo who died in the very year that Newton was born. Galileo Galilei is considered to be the father of modern experimental physics. In his study of moving bodies, he stated informally the principles later embodied in Newton's first two laws of motion. Galileo was the first person to use the telescope to study the heavens. His observations proved (in conflict with religious doctrine) that the solar system is heliocentric. For this he was convicted of heresy by the Inquisition in Rome, ordered to recant and placed under house arrest for the remainder of his life. (The Pope recently issued an apology to Galileo.) His work paved the way for Newton and others to develop the laws of motion and formulate the mathematics of orbits and trajectories.

Galileo had asserted that all objects accelerate to earth at a uniform rate and this rate is the same for all objects no matter what their mass. How might Galileo have discovered this principle?

The following table shows the total distance covered after t seconds (for $0 \le t \le 9$) by an object falling in a vaccuum.

Distance in Meters traveled by a Falling Object in t Seconds										
Time (s)	0	1	2	3	4	5	6	7	8	9
Distance (m)	0	4.9	19.6	44.1	78.4	122.5	176.4	240.1	313.6	396.9

If Galileo were correct, then the average acceleration over each one-second interval should be the same for all intervals. To determine the average acceleration, we create a table showing the average velocity over each one-second interval by subtracting consecutive distances in the table above. We obtain the following table.

Average Velocity of a Falling Object in the tth Second										
Time (s)	0	1	2	3	4	5	6	7	8	9
Av. Velocity (m/s)	0	4.9	14.7	24.5	34.3	44.1	53.9	63.7	73.5	83.3

Subtraction of average velocities for consecutive intervals yields the following table.

Average Acceleration of a Falling Object in the tth Second										
Time (s)	0	1	2	3	4	5	6	7	8	9
Av. Acceleration (m/s²)	9.8	9.8	9.8	9.8	9.8	9.8	9.8	9.8	9.8	9.8

We observe that the object accelerated at a uniform rate of 9.8 m/s^2. Galileo was correct.

WORKED EXAMPLE 1

a) Using the fact that a free falling object has uniform acceleration of 9.8 m/s^2, express the distance s traveled by the object as a function of the time of falling t.

b) Graph the distance vs. time curve and find the distance traveled in the interval $2.8 \le t \le 9.2$. Find the average velocity of the object during this time interval using the distance-time graph.

SOLUTION

a) Acceleration is the rate of change of velocity. Since the acceleration is fixed at 9.8 m/s^2, the velocity is increasing at the constant rate of 9.8 m/s^2. Therefore, the velocity after t seconds is $9.8t$ meters / second. This is shown in the diagram.

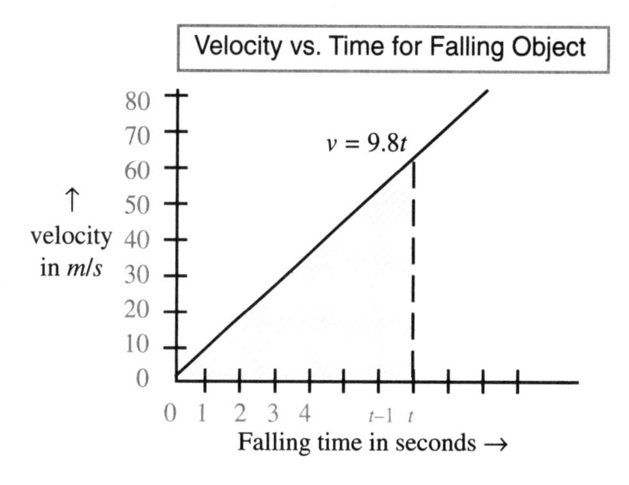

The distance traveled by an object during any time interval is its average velocity (in *meters per second*) during that time interval, times the number of *seconds* in the time interval. Therefore, the distance traveled by the object in the first t seconds of free fall is:

$$\underbrace{\frac{9.8t - 0}{2}}_{\text{average velocity}} \times \underbrace{t}_{\text{time}} \quad \text{or} \quad 4.9t^2$$

The distance traveled in the first t seconds of free fall is represented by the area under the velocity vs. time curve.

We observe that the total distance $s(t)$ is given by $s(t) = 4.9t^2$. This is also represented by the area under the velocity-time graph in the interval $[0, t]$. Galileo used this idea in 1638.

b) To graph this function, we define $y1(x) = 4.9x^2$ in the Y = Editor. Then we graph this function, using the window settings: $-1 \le x \le 11$; $-1 \le y \le 500$. To find the value of the distance traveled when $t = 2.8$ s, (i.e. $x = 2.8$), we select **Value** from the F5 menu, by pressing: **F5** **ENTER**.

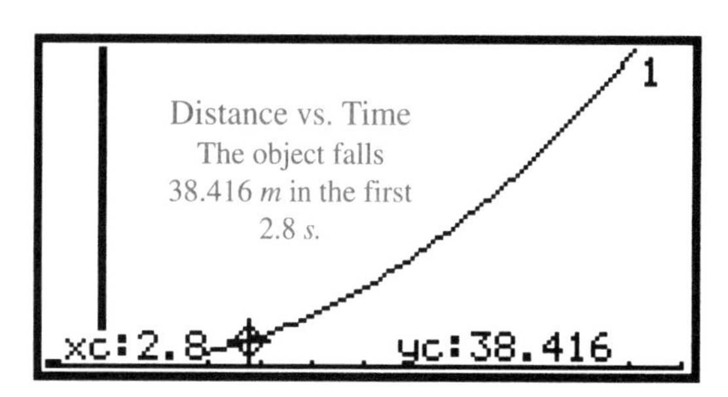

In response to the prompt, **Eval x = ?**, we enter **2.8**. The display gives the corresponding distance, 38.416 m.

To repeat this procedure for $t = 9.2$, press: **F5** **ENTER**.
In response to the prompt, **Eval x = ?**, we enter **9.2**.
The display yields the corresponding distance, 414.736 m.

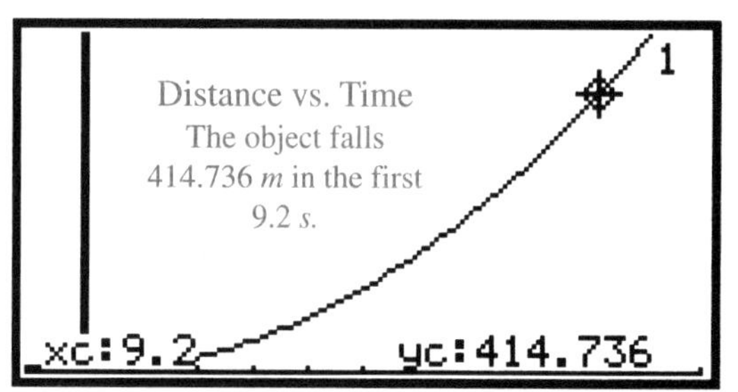

The change in distance fallen between $t = 2.8$ s and $t = 9.2$ s is Δs, where $\Delta s = 414.736 - 38.416 = 376.32$ m.

$$\text{Average velocity} = \frac{\Delta s}{\Delta t} \begin{matrix}\leftarrow \text{change in distance} \\ \leftarrow \text{change in time}\end{matrix}$$

$$= \frac{376.32}{6.4} \quad \text{or } 58.8 \; m/s$$

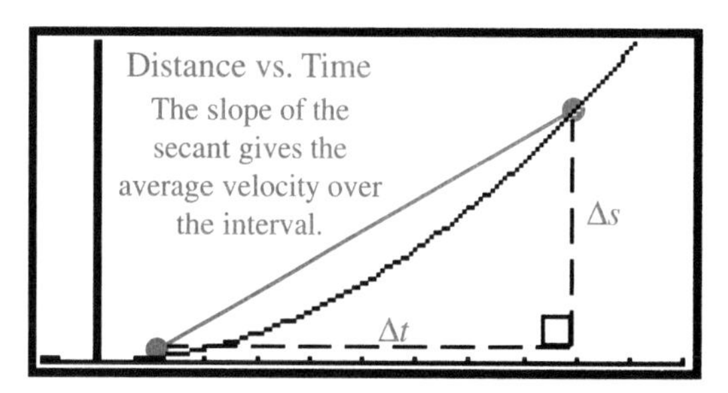

We observe that the average velocity over any time interval is just the slope of the corresponding secant on the distance-time graph.

Worked Examples

In *Worked Example* 1, we were able to derive an equation for the distance as a function of time in the special case when acceleration is a constant. The following example shows how we can determine an approximate distance-time equation when we have a table of distances and corresponding times.

Worked Example 2

This table gives the distances traveled by the 1993 Dodge VIPER RT/10 accelerating from a standing start during a power test reported by *Road & Track* magazine.[1]

Viper Distance vs. Time from Standing Start										
Time in *s*	0	1	2	3	4	5	6	7	8	9
Distance in *m*	0	4.3	14.4	30.1	50.7	76.0	105.5	138.7	175.4	214.9

Express the distance traveled as a function of time and determine the one-second time interval in which the VIPER reached the highest velocity .

VIPER RT/10 / photo courtesy of Chrysler Canada Ltd.

Solution

To find an equation to fit these data as closely as possible, we will enter the data in a matrix variable called VIPER, and construct a scatterplot. We will then find the equation of the cubic curve of best fit and use this as our distance-time function. To access the Data/Matrix Editor, we press:

Then we type the name VIPER and press ENTER *twice*. A blank matrix will appear. We enter the times 0, 1, 2, … , 9 into column c1 of the matrix and the corresponding distances from the table into column c2. We obtain a display like the one shown on the right.

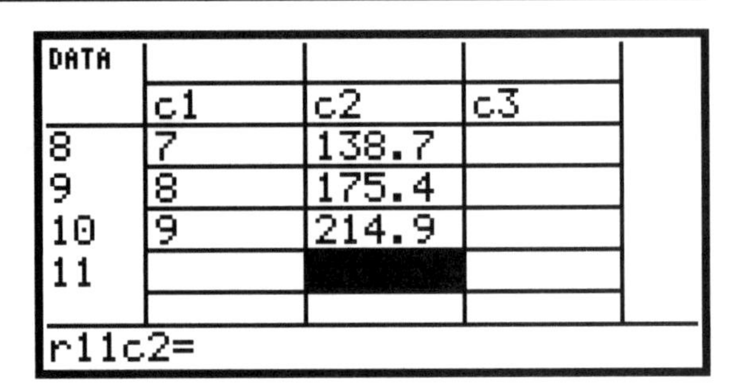

To find the cubic curve of best fit we press F5 and select from the menus, the items shown in this display.

main\viper Calculate
Calculation Type....... CubicReg →
x................................ c1
y................................ c2
Store RegEQ to.......... y1(x) →
Freq and Categories? NO →

After selecting y1(x), we press ENTER *twice*. The equation of the cubic regression curve is presented. We press ENTER . To create the scatterplot, we press F2, then F1 and select Scatter opposite Plot Type. We also enter c1 and c2 for x and y respectively. Pressing ENTER twice returns us to our matrix, VIPER. We then press ◆ [GRAPH] to obtain the display shown here. The distance-time graph is concave upward, showing steadily increasing slope; i.e., the velocity is steadily increasing, reaching its maximum in the final seconds.

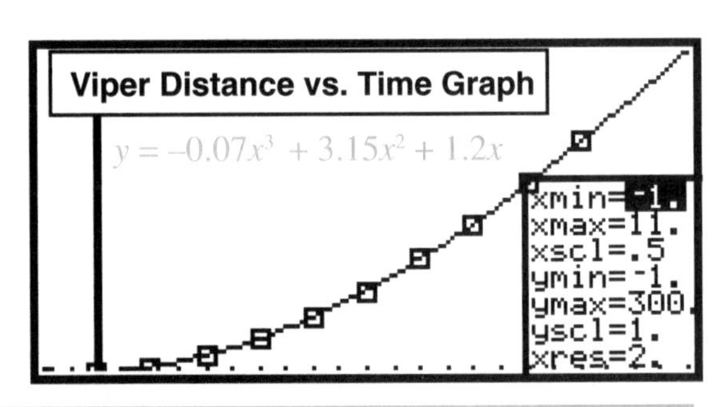

[1] The data for this example derives from an exploration in Ray Nowak's excellent *Mathematical Peepshows* also available through this publisher.

Exercises & Investigations

1. a) State a mathematical relationship relating Δs, Δt and v_{av} given that Δs is the change in distance traveled by a projectile during a time interval Δt, and v_{av} is its average velocity during that time interval.

b) Describe how you can determine v_{av} for any time interval using a distance-time graph.

2. a) State a mathematical relationship relating Δv, Δt and a_{av} given that Δv is the change in velocity of a projectile during a time interval Δt, and a_{av} is its average velocity during that time interval.

b) Describe how you can determine a_{av} for any time interval using a velocity-time graph.

c) Using the velocity-time graph in *Worked Example* 1, state the average velocity over the time interval,

$$2.8 \le t \le 9.2.$$

3. If an object moves so that the acceleration is constant, does this mean that the velocity is a linear function of time? Explain why or why not.

4. a) Suppose a cannon ball were dropped from the top of the leaning tower of Pisa (a height of 56 m). How long would it take for it to reach the ground, if we ignore the effect of wind resistance?

b) The cannon ball would have fallen 44.1 m after 3 seconds and 78.4 m after 4 seconds. Would the distance fallen after 3.5 seconds be the average of these two distances? Explain why or why not.

5. Graph the distance-time equation given in *Worked Example* 2 for the VIPER in the time interval $0 \le t \le 15$.

a) Use the **Value** command on the F5 menu of your graphing screen to determine the distances traveled by the VIPER at $t = 5.5$ and $t = 7.5$ seconds.

b) Calculate the average velocity of the Viper in the interval $5.5 \le t \le 7.5$.

c) Repeat the procedures in parts a) and b) to obtain the average velocity of the VIPER in the interval $5.5 \le t \le 5.6$. Consider this value an estimate of the velocity of the VIPER at $t = 5.5$ seconds.

d) Estimate the velocity of the VIPER at $t = 5.5$ seconds by finding its average velocity over the interval $5.50 \le t \le 5.51$. Compare this estimate with your estimate in part c).

6. Since there is no atmosphere on the moon, a golf ball hit with the impact of a 300-yard drive, would travel according to the equations: $x = 40t$; $y = 40t - 0.9t^2$ where x and y are respectively the horizontal and vertical coordinates in yards, of the golf ball t seconds after impact. Estimate, in yards per second, the vertical velocity of the golf ball 7 seconds after impact.

7. The vertical distance s traveled by an object in free fall for t seconds, is given by $s = 4.9t^2$.

a) Write an expression for $s + \Delta s$, the distance traveled by the object in the first $t + \Delta t$ seconds.

b) Use your answer in part a) to help you write (as a function of t) an expression for Δs, the distance fallen in the interval from t to $t + \Delta t$.

c) Write an expression in terms of t, and Δt for the average velocity during the time interval Δt.

d) Evaluate $\lim\limits_{\Delta t \to 0} \dfrac{\Delta s}{\Delta t}$.

e) Use your answer to part d) to estimate the velocity of an object in free fall, 4.5 seconds after it has been released.

8. Expand each expression and then evaluate the limit (without using your TI-89).

a) $\lim\limits_{\Delta t \to 0} \dfrac{(t+\Delta t)^2 - t^2}{\Delta t}$ b) $\lim\limits_{\Delta t \to 0} \dfrac{(t+\Delta t)^3 - t^3}{\Delta t}$

9. In *Worked Example* 2, we saw that the distance y traveled by the VIPER t seconds into the race is given by the equation:

$$y = -0.07t^3 + 3.15t^2 + 1.2t$$

a) Write an expression for $y + \Delta y$, the distance traveled by VIPER in the first $t + \Delta t$ seconds.

b) Use your answer in part a) to help you write (as a function of t) an expression for Δy, the distance traveled in the interval from t to $t + \Delta t$.

c) Write an expression in terms of t, and Δt for the average velocity during the time interval Δt.

d) Evaluate $\lim\limits_{\Delta t \to 0} \dfrac{\Delta y}{\Delta t}$.

e) Use your answer to part d) to estimate the velocity of VIPER after 5.5 seconds.

10. To evaluate $\lim\limits_{\Delta t \to 0} \dfrac{(t+\Delta t)^2 - t^2}{\Delta t}$ on your TI-89, replace Δt with x, then enter the commands as shown in this display.

```
■ lim  ((t + x)^2 - t^2) / x        2·t
  x→0
limit(((t+x)^2-t^2)/x,x,0…
```

Use this procedure to evaluate $\lim\limits_{\Delta t \to 0} \dfrac{(t+\Delta t)^4 - t^4}{\Delta t}$.

Conjecture a formula for $\lim\limits_{\Delta t \to 0} \dfrac{(t+\Delta t)^n - t^n}{\Delta t}$.

Use your TI-89 to check your conjecture.

EXPLORATION 6 THE DERIVATIVE AS A LIMIT

CORBIS-BETTMANN

Isaac Newton 1642 – 1727

Isaac Newton, Karl Gauss, and Archimedes are rated by historians of mathematics as the three greatest mathematicians of all time. Newton also enjoys such status as a scientist.

In 1665-1666, an outbreak of the bubonic plague forced the closing of Cambridge University. Newton, who had been pursuing his studies there, was sent home to the family farm at Woolsthorpe. There, in an 18-month period of intensely creative work, he made several landmark discoveries that changed the course of mathematics and science. The most mathematically significant of these was his discovery of calculus. To analyse continuously changing quantities which he called *fluents*, (functions of time) he expressed the rate of change of these fluents in terms of other "derived" fluents which he called *fluxions* (and which we now call *derivatives*). Fluents and the techniques for calculating their fluxions (and inverse fluxions) became the basis for a new branch of mathematics which we now call *calculus.* Using calculus, scientists were suddenly able to model in mathematical terms, continuously changing quantities in nature. However, Isaac Newton did not publish his findings in 1666, because an earlier publication on the nature of light had generated such criticism, that Newton vowed never to publish again. Later in life, when a controversy erupted over precedence in the invention of calculus, Newton came to regret his reluctance to publish. Although his discovery of calculus was known to a small cadré of Newton's inner circle in the late 1660's, there was no widely published version of his work until 1704.

CORBIS-BETTMANN

Gottfried Leibniz 1646 – 1716

Gottfried Leibniz, described by historians of mathematics as a "universal genius", was the son of a professor of moral philosophy. A child prodigy, Gottfried devoured all manner of books on matters mathematical and philosophical, graduating with a doctorate at age 20. His early work in the late 1660's was the quest for a universal language (*characteristica universalis*) and an algebra of reasoning (*calculus ratiocinator*) which would reduce moral questions to a sequence of formal computations. This work anticipated, by a century, the subsequent development of symbolic logic.

While in Paris in 1672, Leibniz befriended the great Dutch scientist, Christiaan Huygens, who introduced him to the current frontiers of mathematics. Initial interest in summing infinite series, led Leibniz to investigations involving tangents to curves and areas bounded by them. By 1676, Gottfried Leibniz had discovered the fundamental principles of calculus which Newton had discovered a decade earlier, but not yet published. In 1684, Leibniz published "his" calculus in the journal *Acta Eruditorum* of which he was the editor. This triggered a maelstrom of controversy — the English claiming that he plagiarized Newton's calculus and the Continental Europeans claiming independent discovery. Now with the perspective of hindsight, the historians of mathematics generally attribute credit to both men as independent co-inventors of calculus.

Newton's first important mathematical discovery came during the 1665-1666 hiatus from his formal studies. During those precious months at Woolsthorpe, Newton discovered the Binomial Theorem for fractional exponents. In modern notation, the binomial theorem for fractional exponents is the expansion of $(1 + x)^{m/n}$ as an infinite power series.

The Binomial Theorem for Fractional Exponents

$$(1+x)^{\frac{m}{n}} = 1 + \frac{m}{n}x + \frac{\left(\frac{m}{n}\right)\left(\frac{m}{n}-1\right)}{2!}x^2 + \frac{\left(\frac{m}{n}\right)\left(\frac{m}{n}-1\right)\left(\frac{m}{n}-2\right)}{3!}x^3 + \ldots$$

The Binomial Theorem enabled Newton to compute $f(x + \Delta x) - f(x)$ for algebraic functions of x and thus calculate their derivatives.

WORKED EXAMPLE 1

a) The distance y meters traveled by a rocket in the first x seconds after launch ($x < 120$) is given by

$$y = 0.08x^3 - 1.5x^2 + 24x$$

Graph the distance vs. time equation and find the distance traveled in the first minute.

b) Write an expression for the distance traveled in the first a seconds.
Write an expression for the average velocity over the time interval $a \le x \le a + \Delta x$ where $\Delta x < 1$ and a is any constant such that $0 \le a \le 120$.

c) Evaluate the limit of the average velocity over the interval $a \le x \le a + \Delta x$ as $\Delta x \to 0$, to obtain the instantaneous velocity when $x = a$ seconds.

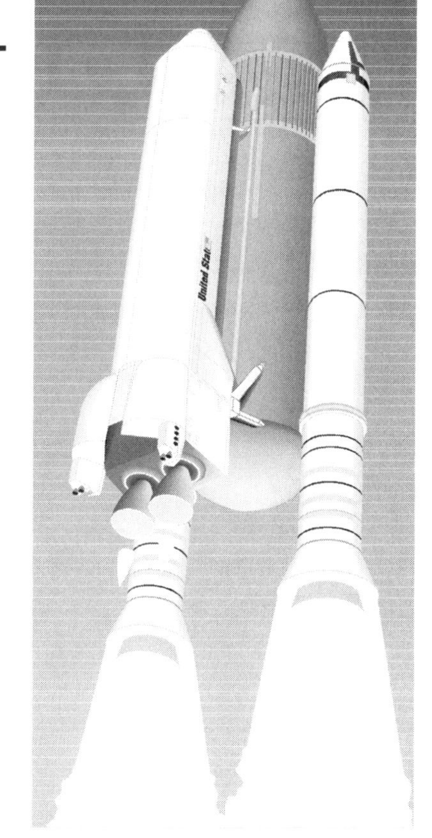

SOLUTION

a) To graph this equation, we define:

$$y1(x) = 0.08x^3 - 1.5x^2 + 24x$$

and we set the window variables as shown in the display.

```
xmin=-10.
xmax=120.
xscl=1.
ymin=-10.
ymax=120000.
yscl=1.
xres=2.
```

To find the distance traveled in the first minute; i.e., 60 seconds, we can enter $y1(60)$ on the command line, or select **Value** from the F5 menu of the graph screen and enter 60 as in *Worked Example* 1 on page 27. We obtain $y1(60) = 13\ 320$ using either method.

b) The distance traveled in the first a seconds is:

$$y1(a) = 0.08a^3 - 1.5a^2 + 24a$$

The average velocity over the interval $a \le x \le a + \Delta x$ is represented on the graph by $\Delta y/\Delta x$; that is, the slope of the secant. PQ

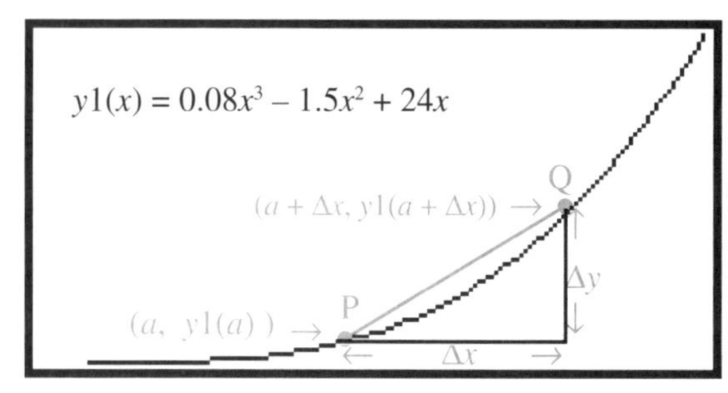

$$\frac{\Delta y}{\Delta x} = \frac{y1(a+\Delta x) - y1(a)}{\Delta x}$$

$$= \frac{[0.08(a+\Delta x)^3 - 1.5(a+\Delta x)^2 + 24(a+\Delta x)] - [0.08a^3 - 1.5a^2 + 24(a)]}{\Delta x}$$

$$= \frac{0.08[3a^2\Delta x + 3a(\Delta x)^2 + (\Delta x)^3] - 1.5[2a\Delta x + (\Delta x)^2] + 24\Delta x}{\Delta x}$$

$$= 0.08(3a^2) - 1.5(2a) + 24 + \Delta x[0.08(3a) - 1.5 + 0.08\Delta x]$$

c) As $\Delta x \to 0$, the point Q slides along the curve toward P and the slope of the secant PQ becomes the slope of the tangent at P. On evaluating the limit as $\Delta x \to 0$, we find

$$\lim_{\Delta x \to 0} \frac{\Delta y}{\Delta x} = 0.08(3a^2) - 1.5(2a) + 24$$

or $0.24a^2 - 3a + 24$ as shown on the display.

Note: To get Δx on your TI-89 press:

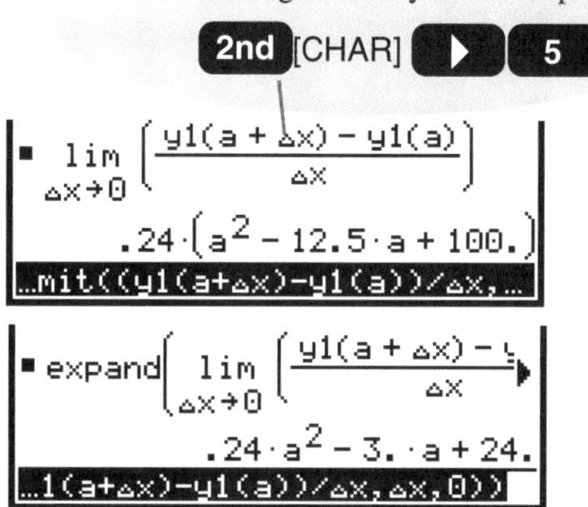

WORKED EXAMPLES

The method used in *Worked Example* 1 is the basis of differential calculus. That is, to find the instantaneous velocity at a point $x = a$, we take the limit of the average velocities over shorter and shorter intervals $a \le x \le a + \Delta x$. Since the instantaneous velocity at $x = a$ is the slope of the tangent to the distance-time graph at $x = a$, this is tantamount to finding the slope of the tangent by taking the limit of the slopes of the secants joining $(a, f(a))$ to $(a + \Delta x, f(a + \Delta x))$ as $\Delta x \to 0$. The expression $0.24a^2 - 3a + 24$, is called the *derivative of* $y1(x)$ *at* $x = a$.

The calculations in *Worked Example* 1 were rather tedious. To simplify the process, we formalize the definition of the derivative.

> **Definition:** The derivative of a function $f(x)$ at $x = a$, denoted by $f'(a)$ is:
>
> $$f'(a) = \lim_{\Delta x \to 0} \frac{f(a+\Delta x) - f(a)}{\Delta x}$$

The foregoing definition assumes that the limit exists. If the limit does not exist, the function is said to have no derivative at $x = a$.

WORKED EXAMPLE 2

a) Find expressions for the derivative of each function $f(x)$ at $x = a$.

i) $f(x) = kx$ ii) $f(x) = kx^2$ iii) $f(x) = kx^3$

where k is any constant.

b) Use your results in part a) to calculate the derivative of the function $y1(x)$ at $x = a$ where $y1(x)$ is the function defined in *Worked Example* 1.

SOLUTION

a) For each function, we construct the quotient $\frac{f(a+\Delta x) - f(a)}{\Delta x}$ and then take the limit as $\Delta x \to 0$.

i) $$\lim_{\Delta x \to 0} \frac{f(a+\Delta x) - f(a)}{\Delta x}$$
$$= \lim_{\Delta x \to 0} \frac{k(a+\Delta x) - k(a)}{\Delta x}$$
$$= \lim_{\Delta x \to 0} \frac{k\Delta x}{\Delta x}$$
$$= k$$

ii) $$\lim_{\Delta x \to 0} \frac{k(a+\Delta x)^2 - k(a^2)}{\Delta x}$$
$$= \lim_{\Delta x \to 0} \frac{k\left[a^2 + 2a\Delta x + (\Delta x)^2\right] - k(a^2)}{\Delta x}$$
$$= \lim_{\Delta x \to 0} \frac{k\left[2a\Delta x + (\Delta x)^2\right]}{\Delta x}$$
$$= 2ak$$

iii) $$\lim_{\Delta x \to 0} \frac{k(a+\Delta x)^3 - k(a^3)}{\Delta x}$$
$$= \lim_{\Delta x \to 0} \frac{k\left[a^3 + 3a^2\Delta x + 3a(\Delta x)^2 + (\Delta x)^3\right] - k(a^3)}{\Delta x}$$
$$= \lim_{\Delta x \to 0} \frac{k\left[3a^2\Delta x + 3a(\Delta x)^2 + (\Delta x)^3\right]}{\Delta x}$$
$$= 3a^2k$$

b) In exercise ❷, you will prove that the derivative of a sum of two or more functions is the sum of the derivatives. That is, if $f_1(x)$ and $f_2(x)$ are two functions, then $(f_1 + f_2)'(a) = f_1'(a) + f_2'(a)$. Therefore, setting $f_1(x) = 0.08x^3$, $f_2(x) = -1.5x^2$ and $f_3(x) = 24x$;

$y1'(a) = 3(0.08)a^2 - 2(1.5)a + 24$,

or $y1'(a) = 0.24a^2 - 3a + 24$, as before.

To obtain the derivative on your TI-89, press: F3 ENTER

Then enter the commands as shown on the command line of the display. You obtain the derivative as shown.

NOTE: THE PROCESS OF FINDING THE DERIVATIVE OF A FUNCTION IS CALLED DIFFERENTIATION.

```
▪ d/da(.08·a^3 - 1.5·a^2 + 24·a)
                .24·a^2 - 3.·a + 24
d(0.08a^3-1.5a^2+24a,a)
```

Exercises & Investigations

❶. a) State in your own words what is meant by the *derivative of a function* $f(x)$ at $x = a$.

b) What does the derivative of $f(x)$ at $x = a$ tell you about the graph of $f(x)$ at $x = a$?

❷. Suppose $f_1(x)$ and $f_2(x)$ are two functions which have a derivative at $x = a$. Use the definition of the derivative to prove that the derivative of $f_1(x) \pm f_2(x)$ at $x = a$ is

$$f_1'(x) \pm f_2'(x).$$

(Assume that the limit of a sum is the sum of the limits.)

❸. Find the derivative of each function at $x = 3$.

a) $f(x) = 4x^2 - 7$ b) $f(x) = 5x^3 + 6x$

c) $f(x) = 2x^3 + 3x^2 - 7$ d) $f(x) = 8x^3 + 4x - 9$

❹. For each of the functions in exercise ❸, find the equation of the tangent to the curve $y = f(x)$ at $x = 3$.

❺. In *Worked Example* 2 p. 28, we saw that the distance y traveled by the VIPER sportscar t seconds into the race is given by the equation:

$$y = -0.07t^3 + 3.15t^2 + 1.2t$$

Calculate the velocity of the VIPER at $t = 5.5$ seconds. Compare with your answer to exercise ❾e) on page 29.

❻. The TI-89 has command **Tangent** on the F5 menu of the graph display. To find the tangent to the graph of $f(x) = 2x^3 + 7x^2 - 4$, at $x = 4$, we graph $f(x)$ in the window:

$$-10 \le x \le 10; \ -1000 \le y \le 1000$$

Then we press: **F5** **alpha** **A** to obtain a display like this:

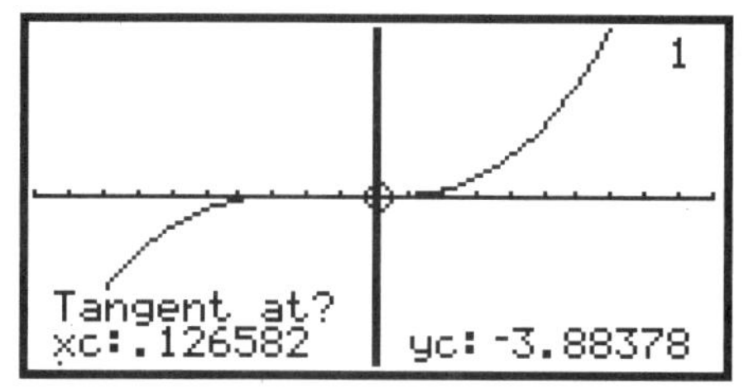

In response to the prompt **Tangent at?**, we enter 4. After pressing **ENTER** we obtain this display.

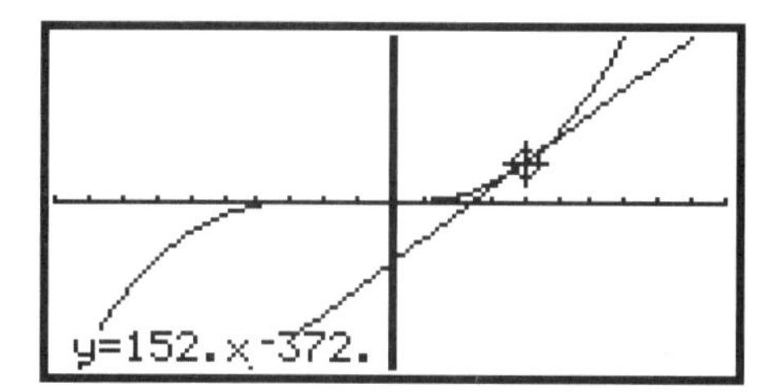

The display gives the equation of the tangent at $x = 4$; that is,

$$y = 152x - 372$$

Use the **Tangent** command on your TI-89 to verify your answers in exercise ❹.

❼. In our definition of the derivative of $f(x)$ at $x = a$, we took the limit over the interval $a \le x \le a + \Delta x$ as $\Delta x \to 0$. Alternatively, we could have defined the derivative of $f(x)$ at $x = a$ as the limit over the interval $a - \Delta x \le x \le a + \Delta x$ as $\Delta x \to 0$. This is called the *central difference quotient*.

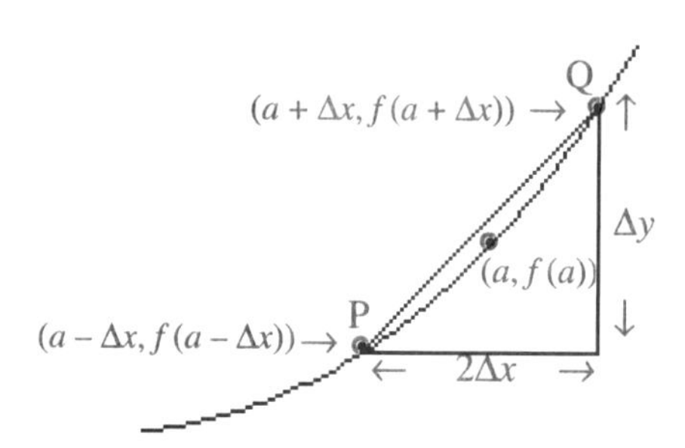

In formal terms, the difference quotient definition of the derivative of $f(x)$ at $x = a$ is:

$$f'(a) = \lim_{\Delta x \to 0} \frac{f(a+\Delta x) - f(a-\Delta x)}{2\Delta x}$$

a) Explain why you might expect the difference quotient definition of $f'(a)$ to be equivalent to the definition of $f'(a)$ given on page 32.

b) The calculus menu (F3) of your TI-89 has the command **nDeriv(** (short for "numerical derivative"). For any function $f(x)$, **nDeriv(*f*(*x*), *x*, *h*)** yields:

$$\frac{f(a+h) - f(a-h)}{2h}$$ ←Note that Δx is replaced by h.

This command automatically forms the different quotient for us. To access **nDeriv(** for $f(x)$, press: **F3** **alpha** **A** . Then enter $f(x)$, x, and h as in the display.

```
■ nDeriv(f(x), x, h)
          -(f(x - h) - f(x + h))
          ----------------------
                  2·h
nDeriv(f(x),x,h)
```

In the command line of the home screen, replace $f(x)$ by x^m ($m > 0$). Then take the limit as $h \to 0$, as shown in the display.

```
limit(nDeriv(x^m,x,h),h,0)
```

c) Write your answer in simplest form. Use your answer in part b) to conjecture a formula for $f'(a)$ when $f(x) = x^m$. Apply your formula to the functions in *Worked Example* 2 on page 32. Compare with the answers given there.

d) To obtain the derivative of $f(x) = x^m$ at any point x, press: **F3** **ENTER** and enter the commands as shown below.

```
d(x^m,x)
```

e) Use the binomial theorem given on page 30 to expand

$$\frac{(x+\Delta x)^m - (x-\Delta x)^m}{2\Delta x}$$ where m is a positive integer

Then take the limit as $\Delta x \to 0$ to find $f'(x)$ for $f(x) = x^m$.

WHO HIT THE LONGEST HOME RUN IN MAJOR LEAGUE HISTORY?

Mickey Mantle 1931-1996

CORBIS-BETTMANN

On September 10, 1960, Mickey Mantle hit the longest home run ever recorded in regular-season major league baseball. In a game between the New York Yankees and the Detroit Tigers at Briggs Stadium in Detroit, he sent the ball into a parabolic orbit. The trajectory of the ball is given by the equation

$$y = 0.9x - 0.0014x^2$$

where x represents the horizontal distance (in feet) and y the vertical distance (in feet) of the ball from home plate. During his career with the New York Yankees (1951-68), Mantle hit a total of 18 home runs in World Series play. This is still a major league record.

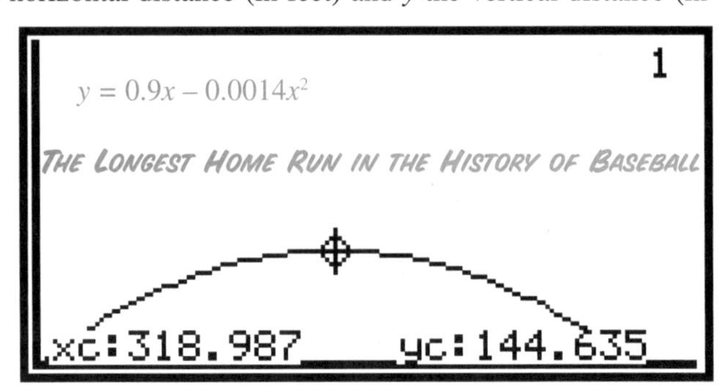

WORKED EXAMPLE 1

a) Graph the trajectory of baseball's longest home run.
b) Determine the maximum height reached by the ball.
c) Determine how far the ball landed from home plate.
d) Determine the angle at which the ball left Mickey Mantle's bat.

SOLUTION

a) We define $y1(x) = 0.9x - 0.0014x^2$ and set the window variables to $0 \le x \le 700$; $0 \le y \le 405$. When we graph $y1(x)$, we obtain the display above.

b) To approximate the maximum height of the ball we can trace (F3) along the curve to the point (318.98…, 144.63…) shown in the display. This tells us that the maximum height is about 145 feet. For a better estimate of the maximum, we press **F5** **4** obtaining the prompt **Lower Bound?** We then trace backwards along the curve to any point left of the maximum and press **ENTER**. This yields the prompt **Upper Bound?**. We then trace to the right of the maximum and press **ENTER**. The display reveals that the maximum in the interval defined by the lower and upper bounds is 144.64….confirming that the maximum height is about 145 feet.

Alternatively, we can actually *calculate* the maximum height by observing that the height is a maximum when the slope of the trajectory is 0; that is, when the derivative of $y1(x)$ is 0. We define $f(x) = 0.9x - 0.0014x^2$, so $f'(x) = 0.9 - 2(0.0014)x$ Solving $f'(x) = 0$, yields $x = 0.9/(0.0028)$ or 321.428…. Evaluating $y1(321.428)$ yields 144.64… as above.

c) To approximate the length of the home run, we can trace along the graph to (641.17…, 1.50…) where the trajectory appears to intersect the x-axis. This yields a distance of 641 feet. To actually *calculate* the length, we can solve $y1(x) = 0$ using the formula for the roots of a quadratic equation, or by entering solve **(y1(x) = 0, x)** on the command line. In either case we obtain $x = 642.857…$, indicating that the horizontal distance traveled by the ball was about 643 feet.

d) The slope of the trajectory at $x = 0$ is $f'(0)$, i.e., the tangent of angle at which the ball left Mickey Mantle's bat. But $f'(0) = 0.9$. The angle of inclination at impact was $\tan^{-1}(0.9) \approx 42°$.

WORKED EXAMPLES

The recent realization that the world's natural resources are in limited supply has made conservation an international priority of the 1990's. One of the world's most valued commodities which is now in short supply is tin.The worked example below is typical of many types of problems that applied mathemati - cians and engineers confront in the design of containers and packaging.

WORKED EXAMPLE 2

A cylindrical tin can must be designed to hold 300 *mL* of liquid. What dimensions for the can would require the minimum amount of tin? (Given the dimensions to the nearest hundredth of a centimeter.)

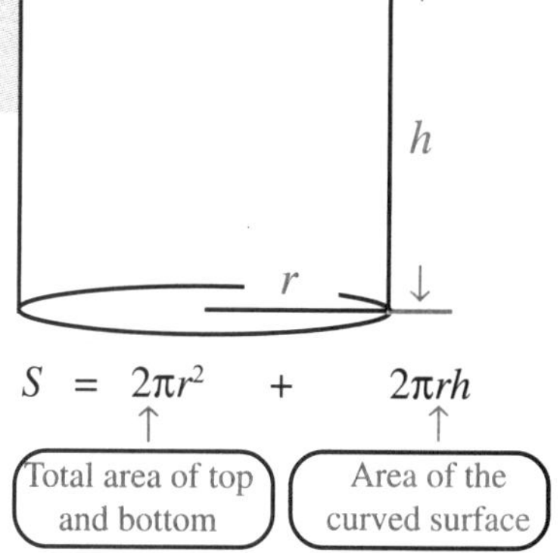

SOLUTION

We assume the tin can is to be a right circular cylinder (for stacking purposes). We denote its height and the radius of its base in centimeters by h and r respectively.

We must find values of r and h which minimize the surface area S while maintaining a volume of 300 cm^3 (Recall 1 $cm^3 = 1\ mL$.) That is, we must minimize S subject to the condition $\pi r^2 h = 300$ where $S = 2\pi r^2 + 2\pi rh$.

Since $\pi r^2 h = 300$, we can substitute $\frac{300}{\pi r^2}$ for h in the formula for S to obtain: $S = 2\pi r^2 + \frac{600}{r}$

To graph S as a function of r on the TI-89, we represent r by the variable x and S by $y1(x)$.

We enter in the Y = Editor: $y1(x) = 2\pi x^2 + \frac{600}{x}$. To determine the appropriate window settings, we press: [◆] [TABLE] and scan the values of $y1(x)$ for $x = 1, 2, \ldots 5$ as shown in the display. We see that $y1(x)$ decreases from $x = 1$ and begins to increase again near $x = 4$. We set the viewing window variables to $1 \le x \le 7$; $200 \le y \le 400$. After graphing $y1(x)$, we press:

x	y1
1.	606.28
2.	325.13
3.	256.55
4.	250.53
5.	277.08

x=5.

[F5] [3] to access **Minimum** on the calculus menu. We then respond to the prompts **Lower Bound?** and **Upper Bound?** as in *Worked Example* 1 and we obtain the display shown below. We observe that the minimum surface area is about 248 cm^2 and this occurs when $r \approx 3.63\ cm$ and $h \approx 7.25\ cm$.(In general, the surface area is a minimum for $h = 2r$.)

To *calculate* the minimum, we observe that the minimum occurs when the slope of the curve is 0. That is, when the derivative of $y1(x)$ is 0. We access **d(** on the calculus menu and enter the commands as shown to obtain this display.

$$\blacksquare \frac{d}{dx}(y1(x)) \qquad 4 \cdot \pi \cdot x - \frac{600}{x^2}$$

```
d(y1(x),x)
```

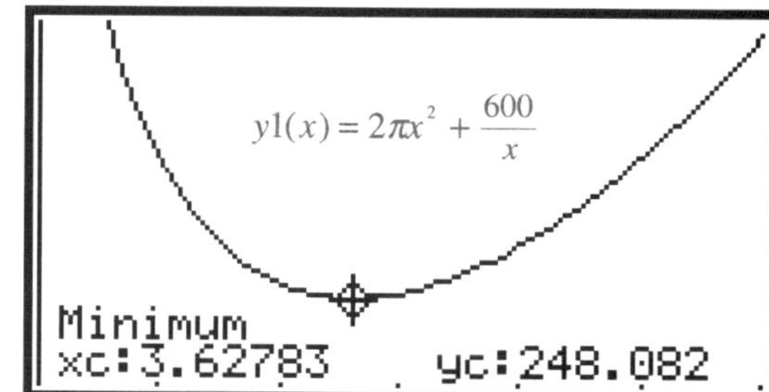

We then set this derivative equal to 0; i.e., $4\pi x - \frac{600}{x^2} = 0$ . .

Solving for x yields $x = \sqrt[3]{\frac{150}{\pi}} = 3.62783167\ldots$

We see that the estimate was very close to the true value.

EXERCISES & INVESTIGATIONS

❶. In *Worked Example* 1 on page 34:
a) What is the name of the curve that describes the trajectory of the baseball?
b) What is the horizontal distance of the ball from home plate when it reaches the highest point in its trajectory?
c) What is the slope of the secant which joins the point on the trajectory at $x = 0$ to the highest point in the trajectory?
d) What is the slope of the tangent to the trajectory at its highest point? Give reasons for your answer.
e) The equation of the trajectory of Mickey Mantle's home run is a mathematical model of the path of the baseball. Do you think this model is accurate? Explain.

❷. A function has a *local maximum at* $x = a$ if and only if $f(x) \leq f(a)$ for all values of x close to a. The value $f(a)$ is said to be a *local maximum.*
a) Define a *local minimum.*
b) Does the function defined by the equation $y = x^3 + 8$, have either a local maximum or a local minimum?

❸. Graph the function $f(x)$ where $f(x) = 2x^3 - 13x^2 - 7x + 1$.
a) Select the Maximum and Minimum commands on the F5 menu of the graph screen to determine all the local maxima and minima.
b) Solve the equation $f'(x) = 0$ to find the local maxima and minima and compare with your answer in part a).

❹. Determine all the points at which each of these functions has a local maximum or minimum.
a) $f(x) = 3x^2 - 7x + 5$
b) $f(x) = x^4 + 2x^3 - 3x^2 - 4x - 4$
c) $f(x) = x^4 - 6x^3 + 12x^2 - 8x - 5$

❺. A local maximum or minimum is called an *extremum.* What is the largest number of extrema which $f(x)$ can have if $f(x)$ is:
a) a linear function? b) a quadratic function?
c) a cubic function? d) a polynomial of degree n?

❻. Write an equation to define a function of x which has a maximum when $x = 0$, and a minimum when $x = 4$. Is the function you have defined the only one satisfying these conditions?

❼. Two sides of a triangular sign are to be each 5 m long. What length of the third side (to the nearest centimeter) would yield the maximum possible area?

❽. Determine the coordinates of the point on the line with equation $y = 3x + 2$, which is closest to the origin, (0, 0).

❾. What point on the parabola with equation $y = x^2 - 4x + 3$ is closest to the origin?

❿. a) Determine the distance between the lines defined by equations $y = 2.7x - 13$ and $y = 2.7x + 16$. (Give your answer to two decimal places.)
b) Check your answer to part a) by finding the point of intersection of the line defined by equation $y = 2.7x + 16$ and the line perpendicular to it and passing through the point (0, – 13).

⓫. In a heavy fog at 7:00 a.m., a cargo ship is 80 km due east of a luxury liner which is sailing due south at a speed of 42 *km/h*. If the cargo ship is sailing due west at 48 *km/h*, at what time will they attain their closest approach? How close will they be?

⓬. The total cost in dollars of producing x grummets is given by the formula $C = 0.35x^2 + 23x + 20$. Each grummet sells for $80.
a) How many grummets should be produced to achieve a maximum profit?
b) How much profit would this generate?

⓭. Ms. Chiu plans to put a rectangular swimming pool in her yard. Since one side of the pool will run along an existing fence it will be necessary for her to fence only three sides. What dimensions of the enclosed rectangle would maximize the contained area if Ms. Chiu has 35 m of fencing?

⓮. The diagram below shows a beam of length L laid across a wall 7 m tall to support the side of a building. The wall is 11 m from the building.

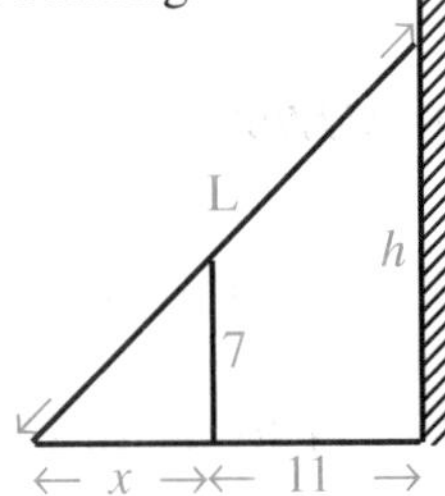

a) Denote the distance between the foot of the beam and the wall by x. Write an equation which relates x and h where h is the height of the upper end of the beam. (Use similar triangles.)
b) Write an equation for the length of the beam L in terms of x and h.
c) Use your equations in parts a) and b) of this exercise to express L in terms of the variable x only.
d) Graph L as a function of x and determine the minimum possible length of the beam.
e) How far is the bottom of the beam from the wall when the beam is the minimum possible length?
f) Does the function of x you graphed have more than one minimum? Intepret each minimum.
g) Is there a maximum length which the beam can have? Explain your answer.

Unit 3

Formal Differentiation

Mathematical Concepts

- derivatives of sums, differences, products & quotients
- Generalized Power Rule for Derivatives
- conditions for maxima & minima
- points of inflection
- definition of *e*
- derivatives of exponential & log functions
- derivatives of hyperbolic functions
- derivatives of trigonometric functions
- L'Hôpital's Rule
- The Chain Rule
- related rates and motion
- parametric equations
- higher order derivatives
- The Prime Number Theorem

TI-89 Commands

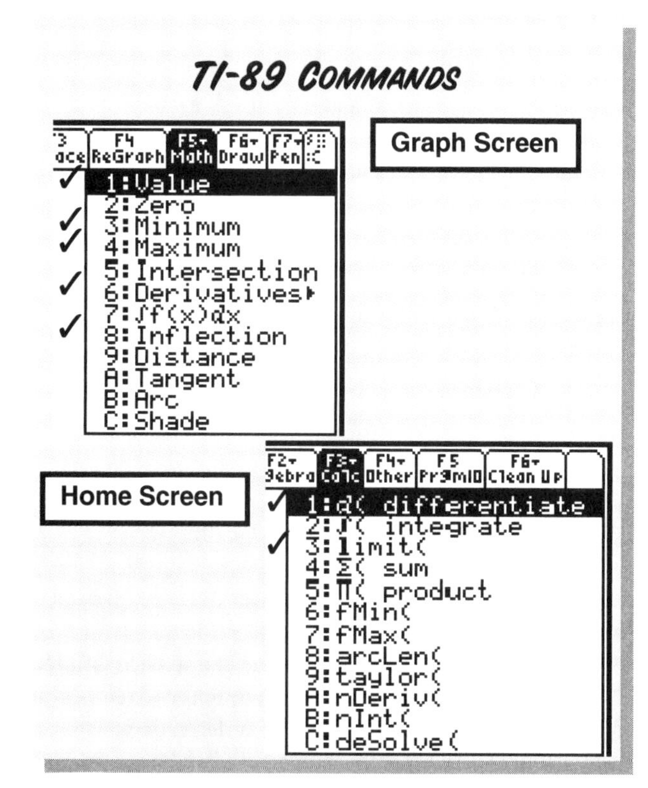

Derivatives of Polynomials & Rational Functions

CORBIS-BETTMANN

Augustin Cauchy 1789 – 1848

CORBIS-BETTMANN

Karl Weierstrass 1815 – 1897

In the next few explorations, we will formalize and extend concepts introduced in the previous unit. While the seventeenth and eighteenth centuries saw the conceptualization of calculus, this new branch of mathematics was not set upon a firm foundation until the nineteenth century. The refinements necessary to render calculus a rigorous and logically defensible discipline were carried out by the French mathematician Augustin Cauchy, Austrian mathematician Karl Weierstrass, and others. Through their contributions, the definitions of limit and continuity were formalized and the tradition of ε–δ proofs emerged. Calculus blossomed into a wider branch of mathematics now called *analysis*, and the concepts of function and space were defined and generalized.

Previously we have used the notation $f'(x)$ to denote the derivative of the function $y = f(x)$. Often we will represent the derivative of $f(x)$ by the symbol $\frac{dy}{dx}$ (read "the derivative of y with respect to x" or simply, "dy by dx". This notation was first introduced by Leibniz. It has the distinct advantage that it is suggestive of the quotient implicit in the definition of the derivative. The significance of this advantage will be particularly evident when we encounter the *chain rule* in *Exploration* 13.

Definition: A function $f(x)$ is said to be *differentiable over the interval* $[a, b]$ if and only if $\frac{dy}{dx}$ exists at every point in that interval.

Furthermore we say that $f(x)$ is *differentiable* if it is differentiable over the entire real line.

In exercise ❽ you will prove the following theorem for differentiable functions.

Theorem

If $f(x)$ and $g(x)$ are differentiable functions, then:

$$\frac{d[f(x) \pm g(x)]}{dx} = \frac{df(x)}{dx} \pm \frac{dg(x)}{dx}$$

This theorem shows that the derivative of a sum (or difference) of two functions is the sum (or difference) of their derivatives. This prompts us to conjecture that the derivative of the product of two functions is the product of their derivatives. For a while, Leibniz thought this was so. Later he discovered that this conjecture is false, as the following theorem shows.

WORKED EXAMPLES

Theorem

If $f(x)$ and $g(x)$ are differentiable functions, then:

$$\frac{d[f(x)g(x)]}{dx} = f(x)\frac{dg(x)}{dx} + g(x)\frac{df(x)}{dx}$$

VISUALIZE

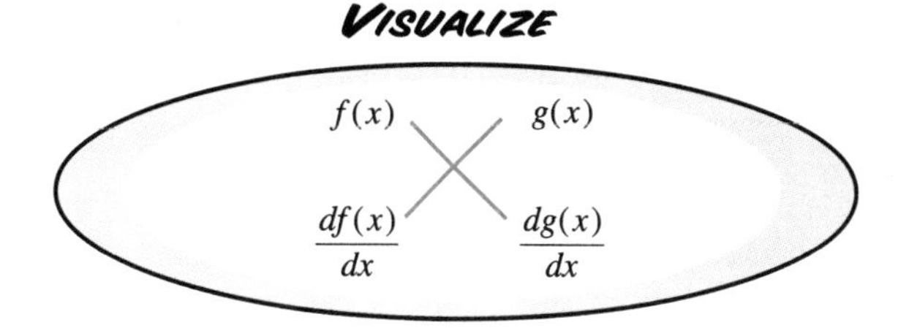

Proof

$$\frac{d[f(x)g(x)]}{dx} = \lim_{\Delta x \to 0} \frac{f(x+\Delta x)\,g(x+\Delta x) - f(x)\,g(x)}{\Delta x}$$ ← by the definition of the derivative of $f(x)g(x)$

adding and subtracting $f(x+\Delta x)g(x)$

$$= \lim_{\Delta x \to 0} \frac{f(x+\Delta x)\,g(x+\Delta x) - f(x+\Delta x)g(x) + f(x+\Delta x)g(x) - f(x)\,g(x)}{\Delta x}$$

$$= \lim_{\Delta x \to 0} \frac{f(x+\Delta x)\,g(x+\Delta x) - f(x+\Delta x)\,g(x)}{\Delta x} + \lim_{\Delta x \to 0} \frac{f(x+\Delta x)\,g(x) - f(x)\,g(x)}{\Delta x}$$

$$= f(x)\frac{dg(x)}{dx} + g(x)\frac{df(x)}{dx}$$

Corollary

If $f(x)$ and $g(x)$ are differentiable functions, then:

$$\frac{d\left[\frac{f(x)}{g(x)}\right]}{dx} = \frac{g(x)\frac{df(x)}{dx} - f(x)\frac{dg(x)}{dx}}{g(x)^2}$$

You will prove this corollary in exercise ❽, by following the procedure in the proof of the theorem above.

Using the foregoing theorem, we can evaluate $\frac{d}{dx}(x^m)$ for positive integral values of m as shown below.

WORKED EXAMPLE 1

Prove: a) $\frac{d}{dx}(x^2) = 2x$ b) $\frac{d}{dx}(x^3) = 3x^2$ c) $\frac{d}{dx}(x^m) = mx^{m-1}$ for positive integral m

SOLUTION

a) Let $f(x) = x$ and $g(x) = x$, then :

$\frac{d}{dx}(f(x)) = 1$ and $\frac{d}{dx}(g(x)) = 1$, so $\frac{d}{dx}(x^2) = \frac{d[f(x)g(x)]}{dx} = f(x)\frac{dg(x)}{dx} + g(x)\frac{df(x)}{dx} = x + x$ or $2x$.

b) Let $f(x) = x^2$ and $g(x) = x$, then :

$\frac{d}{dx}(f(x)) = 2x$ and $\frac{d}{dx}(g(x)) = 1$, so $\frac{d}{dx}(x^3) = \frac{d[f(x)g(x)]}{dx} = f(x)\frac{dg(x)}{dx} + g(x)\frac{df(x)}{dx} = x^2 + x(2x)$ or $3x^2$.

c) To prove the formula true for any positive integer m, we use mathematical induction. We know the theorem is true for $m = 1$. (Why?) Assume the theorem is true for $m - 1$. Then $\frac{d}{dx}(x^{m-1}) = (m-1)x^{m-2}$.

Let $f(x) = x^{m-1}$ and $g(x) = x$, then :

$\frac{d}{dx}(f(x)) = (m-1)x^{m-2}$ and $\frac{d}{dx}(g(x)) = 1$, so $\frac{d}{dx}(x^m) = f(x)\frac{dg(x)}{dx} + g(x)\frac{df(x)}{dx} = x^{m-1}(1) + x(m-1)x^{m-2}$ or mx^{m-1}.

Therefore, the formula is true for *all* positive integers, m.

Worked Examples

In *Worked Example* 1 we have proved the following formula called the *power rule* for differentiating powers of x with respect to x.

Power Rule $\frac{d}{dx}(x^m) = mx^{m-1}$ for positive integral m

We can use the power rule in conjunction with the theorems on pages 38 and 39 to calculate the derivative of any polynomial or rational function.

Worked Example 2

Differentiate $f(x) = 3x^5 - 7x^3 + 2x - 3$ with respect to x:
a) using the power rule b) using your TI-89

Solution

a) $\frac{d}{dx}(3x^5-7x^3+2x-3) = \frac{d}{dx}(3x^5) - \frac{d}{dx}(7x^3) + \frac{d}{dx}(2x) - \frac{d}{dx}(3)$ ← The theorem on page 38 indicates that the derivative of a sum or difference is the sum or difference of the derivatives.

$= 15x^4 \quad - 21x^2 \quad + 2 \quad - 0$ ☚ applying the power rule

$= 15x^4 - 21x^2 + 2$

Note: The derivative of a constant is 0. How would you explain this fact geometrically?

b) Press F3 ENTER. Then enter the polynomial $f(x)$ as shown in the command line of the display. We obtain the answer shown in the display. This verifies part a).

```
▪ d/dx(3·x^5 − 7·x^3 + 2·x − 3)
                     15·x^4 − 21·x^2 + 2
d(3x^5-7x^3+2x-3,x)
```

Worked Example 3

Differentiate $\frac{d}{dx}\left(\frac{3x^5-7x^3+2x-3}{x^2+1}\right)$ with respect to x:
a) using the theorem on page 39 and the power rule b) using your TI-89

Solution

a) Let $f(x) = 3x^5 - 7x^3 + 2x - 3$. Then, from *Worked Example* 2, $\frac{d}{dx}(f(x)) = 15x^4 - 21x^2 + 2$.
Let $g(x) = x^2 + 1$. Then $\frac{d}{dx}(g(x)) = 2x$.

$$\frac{d}{dx}\left(\frac{3x^5-7x^3+2x-3}{x^2+1}\right) = \frac{(3x^5-7x^3+2x-3)(2x)-(15x^4-21x^2+2)(x^2+1)}{(x^2+1)^2}$$ ☚ applying the corollary on page 39

$$= \frac{9x^6+8x^4-23x^2+6x+2}{(x^2+1)^2}$$

b) Press F3 ENTER. Then enter the rational function as shown in the command line of the display. We obtain the answer shown in the display. This verifies the answer obtained in part a)

```
▪ d/dx((3·x^5 − 7·x^3 + 2·x − 3)/(x^2 + 1))
   (9·x^6 + 8·x^4 − 23·x^2 + 6·x + ▸)/(x^2 + 1)^2
...^5-7x^3+2x-3)/(x^2+1),x)
```

EXERCISES & INVESTIGATIONS

❶. Explain why the derivative of the constant function, $f(x) = a$ is everywhere equal to 0.

❷. Prove that $\frac{d}{dx}(kf(x)) = k\frac{d}{dx}(f(x))$ for any constant k.

❸. Differentiate $f(x)$ with respect to x where:

a) $f(x) = 5x^3 - 4x^2 + 3x - 7$
b) $f(x) = 6x^4 - 5x^3 + 2x + 3$
c) $f(x) = 3x^7 - 4x^5 + 7x^3 - 21$

Graph $y = f(x)$ and $y = f'(x)$ in each case, in the window $-10 \le x \le 10$; $-100 \le y \le 100$.

❹. Use your answers in exercise ❸ to find the equation of the tangent line to each graph at the given point.

a) $y = 5x^3 - 4x^2 + 3x - 7$ at (2, 23)
b) $y = 6x^4 - 5x^3 + 2x + 3$ at (–1, 12)
c) $y = 3x^7 - 4x^5 + 7x^3 - 21$ at (1, –15)

❺. Differentiate $f(x)$ with respect to x where:

a) $f(x) = 3x^2(4x^3 - 6x + 4)$
b) $f(x) = (5x^2 - 1)(3x^3 + 2x + 4)$
c) $f(x) = (6x^2 - x + 3)(2x^3 - 6x)$
d) $f(x) = (2x^2 + 1)^3$

Graph $y = f(x)$ and $y = f'(x)$ in each case, in the window $-10 \le x \le 10$; $-100 \le y \le 100$.

❻. Use your answers in exercise ❺ to find the equation of the tangent line to each graph at the given point.

a) $y = 3x^2(4x^3 - 6x + 4)$ at (–1/2, 39/8)
b) $y = (5x^2 - 1)(3x^3 + 2x + 4)$ at (1, 36)
c) $y = (6x^2 - x + 3)(2x^3 - 6x)$ at (2, 100)
d) $y = (2x^2 + 1)^3$ at (0, 1)

❼. Differentiate with respect to x.

a) $\frac{3x}{4x^2-7}$ b) $\frac{5x^2+3x+4}{2x-9}$ c) $\frac{(2x+1)^2}{6x^3-11x^2-7x}$

❽. Prove that if $f(x)$ and $g(x)$ are differentiable functions,

a) $$\frac{d[f(x) \pm g(x)]}{dx} = \frac{df(x)}{dx} \pm \frac{dg(x)}{dx}$$

b) $$\frac{d\left[\frac{f(x)}{g(x)}\right]}{dx} = \frac{g(x)\frac{df(x)}{dx} - f(x)\frac{dg(x)}{dx}}{g(x)^2}$$

❾. Use the rule for differentiating quotients (corollary on page 39) to prove that the power rule, $\frac{d}{dx}(x^m) = mx^{m-1}$ is valid for *all* integral values of m.

❿. In investigation ❾ of Exploration 5 (p. 29), we encountered the equation $y = -0.07t^3 + 3.15t^2 + 1.2t$, which expresses the distance y in meters traveled by the RT/10 VIPER in the first t seconds of a race ($t < 30$).

a) Write an expression for the instantaneous velocity of the VIPER as a function of the time, t.

b) Graph the function obtained in part a) and identify the time at which the VIPER reaches maximum velocity during the first 20 seconds.

c) If acceleration is defined to be the instantaneous rate of change of velocity, write an expression for the acceleration of the VIPER as a function of time. At what time in the first 20 seconds is the acceleration a maximum? Explain what this means.

d) Calculate the instantaneous velocity of the VIPER when $t = 5.5$ seconds. Compare your answer with the answer you obtained to investigation ❾ e) on page 29.

NEWTON'S DERIVATION OF THE POWER RULE

In *Exploration* 6 page 30, we saw the following binomial expansion for rational exponents that Newton discovered during his 1665-1666 stay in Woolsthorpe. In this investigation, we will use this expansion to develop the power rule for integral exponents.

$$(1+x)^{\frac{m}{n}} = 1 + \frac{m}{n}x + \frac{\left(\frac{m}{n}\right)\left(\frac{m}{n}-1\right)}{2!}x^2 + \frac{\left(\frac{m}{n}\right)\left(\frac{m}{n}-1\right)\left(\frac{m}{n}-2\right)}{3!}x^3 + \ldots$$

a) Show that this expansion reduces to the binomial theorem for integral exponents when $n = 1$. Why does the series on the right become finite when m/n is integral?

b) Use the expansion you derived in part a) to obtain an expansion for $\left(\frac{x+\Delta x}{x}\right)^m$.

c) Use the expansion you derived in part b) to obtain an expansion for $(x + \Delta x)^m - x^m$.

d) Use your expansion in part c) to derive an expression for $\frac{d}{dx}(x^m)$

EXPLORATION 9 DERIVATIVES OF ALGEBRAIC FUNCTIONS

In *Exploration* 8, we used mathematical induction to derive the power rule for differentiating positive integral powers of x.

$$\frac{d}{dx}(x^m) = mx^{m-1} \quad \text{for positive integral } m$$

In exercise ❾ on page 41, you showed that this power rule is valid for *all* integral powers of x. Is this rule valid for all rational values of x including positive and negative fractions?

One way to determine whether the power rule applies to rational exponents would be to calculate numerical derivatives of functions such as $y = \sqrt{x}$ and determine whether these values coincide with the function obtained by applying the power rule; i.e., for $f(x) = \sqrt{x}$ the power rule would yield $f'(x) = 1/(2\sqrt{x})$. If the numerical derivatives and the theoretical derivative coincided, then we might conjecture that the power extends to fractional exponents. But how could we prove this conjecture? Certainly we could not use mathematical induction to extend the power rule to the rationals. Why not?

"WHAT AN EGO! ONE DAY IT'S NEWTON'S LAWS OF DYNAMICS, THEN IT'S NEWTON'S THEORY OF GRAVITATION, AND NEWTON'S LAW OF HYDRONAMIC RESISTANCE, AND NEWTON'S THIS AND NEWTON'S THAT."

It is a testimonial to the genius of Isaac Newton that he was able to invent the required tool when he needed it for the development of his calculus. This device was the binomial theorem for rational exponents presented on page 30.

$$(1+x)^{\frac{m}{n}} = 1 + \frac{m}{n}x + \frac{\left(\frac{m}{n}\right)\left(\frac{m}{n}-1\right)}{2!}x^2 + \frac{\left(\frac{m}{n}\right)\left(\frac{m}{n}-1\right)\left(\frac{m}{n}-2\right)}{3!}x^3 + \dots$$

The investigation at the bottom of page 41 guided you through a proof of the power rule for positive integral exponents by setting $n = 1$ in the above expansion. There follows a proof of the power rule for any rational exponent m/n, using this expansion.

$$\frac{d}{dx}\left(x^{\frac{m}{n}}\right) = \lim_{\Delta x \to 0} \frac{(x+\Delta x)^{\frac{m}{n}} - x^{\frac{m}{n}}}{\Delta x}$$

$$= x^{\frac{m}{n}} \lim_{\Delta x \to 0} \frac{\left(1+\frac{\Delta x}{x}\right)^{\frac{m}{n}} - 1}{\Delta x} \quad \leftarrow \text{extracting the factor } x^{m/n}$$

$$= x^{\frac{m}{n}} \lim_{\Delta x \to 0} \frac{\left[1 + \frac{m}{n}\left(\frac{\Delta x}{x}\right) + \frac{\left(\frac{m}{n}\right)\left(\frac{m}{n}-1\right)}{2!}\left(\frac{\Delta x}{x}\right)^2 + \frac{\left(\frac{m}{n}\right)\left(\frac{m}{n}-1\right)\left(\frac{m}{n}-2\right)}{3!}\left(\frac{\Delta x}{x}\right)^3 + \dots\right] - 1}{\Delta x} \quad \leftarrow \text{substituting } \Delta x/x \text{ for } x \text{ in the expansion above}$$

$$= x^{\frac{m}{n}}\left(\frac{m}{n}\left(\frac{1}{x}\right)\right) \quad \leftarrow \text{all terms inside the square brackets above vanish except the first two as } \Delta x \to 0$$

$$= \frac{m}{n}x^{\frac{m}{n}-1} \qquad \text{That is, } \frac{d}{dx}\left(x^{\frac{m}{n}}\right) = \frac{m}{n}x^{\frac{m}{n}-1}$$

This shows that the power rule extends to all rational exponents. We write

$$\boxed{\frac{d}{dx}(x^m) = mx^{m-1} \quad \text{for all rational exponents } m}$$

Worked Example 1

Differentiate with respect to x.

a) $x^{\frac{2}{3}} - x^{-\frac{1}{2}} + x^2 - 2$ b) $\sqrt{x^2+36}$ c) $\sqrt{(15-x)^2+10^2}$

Solution

a) Using the power rule for fractional exponents, we write:

$$\frac{d}{dx}\left(x^{\frac{2}{3}} - x^{-\frac{1}{2}} + x^2 - 2\right) = \frac{2}{3}x^{-\frac{1}{3}} + \frac{1}{2}x^{-\frac{3}{2}} + 2x$$

To check this on your TI-89, press **F3** **ENTER** and enter the algebraic function as shown in the command line of this display.

b) Our power rule applies only to powers of monomials. (We will extend it again in *Exploration* 13.) Therefore, to perform this differentiation, we will differentiate it as a product and use the fact that $\sqrt{x^2+36} \times \sqrt{x^2+36} = x^2+36$.

$$\frac{d}{dx}(x^2+36) = \frac{d}{dx}\left(\left(\sqrt{x^2+36}\right)\left(\sqrt{x^2+36}\right)\right) = \sqrt{x^2+36}\,\frac{d}{dx}\left(\sqrt{x^2+36}\right) + \sqrt{x^2+36}\,\frac{d}{dx}\left(\sqrt{x^2+36}\right)$$

$$= 2\sqrt{x^2+36}\,\frac{d}{dx}\left(\sqrt{x^2+36}\right) \quad \text{and so,}$$

$$\frac{d}{dx}\left(\sqrt{x^2+36}\right) = \frac{1}{2\sqrt{x^2+36}}\,\frac{d}{dx}(x^2+36)$$

$$= \frac{x}{\sqrt{x^2+36}}$$

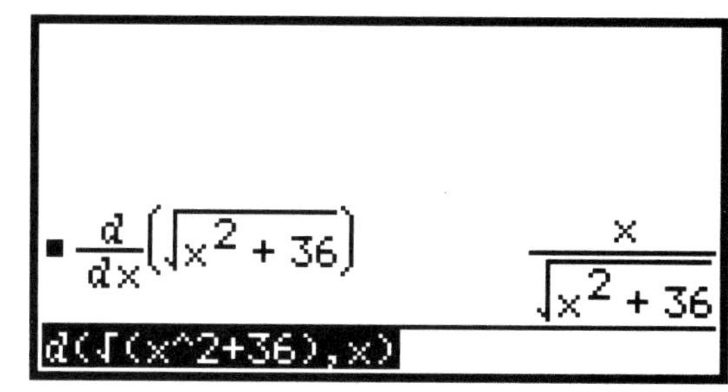

To check this on your TI-89, press **F3** **ENTER** and enter the function as shown in the command line of the display.

c) Proceeding as in part b), we obtain:

$$\frac{d}{dx}\left(\sqrt{(15-x)^2+10^2}\right) = \frac{1}{2\sqrt{(15-x)^2+10^2}}\,\frac{d}{dx}\left((15-x)^2+10^2\right) = \frac{x-15}{\sqrt{(15-x)^2+10^2}}$$

To check this on your TI-89, press **F3** **ENTER** and enter the algebraic function as shown in the command line of the display.

The method in *Worked Example* 1 can be generalized (see investigation 10 p. 45) to prove that for any rational power m of a function $f(x)$:

Generalized Power Rule

$$\frac{d}{dx}\left(f(x)^m\right) = mf(x)^{m-1}\,\frac{d}{dx}\left(f(x)\right) \quad \text{for any rational } m$$

WORKED EXAMPLES

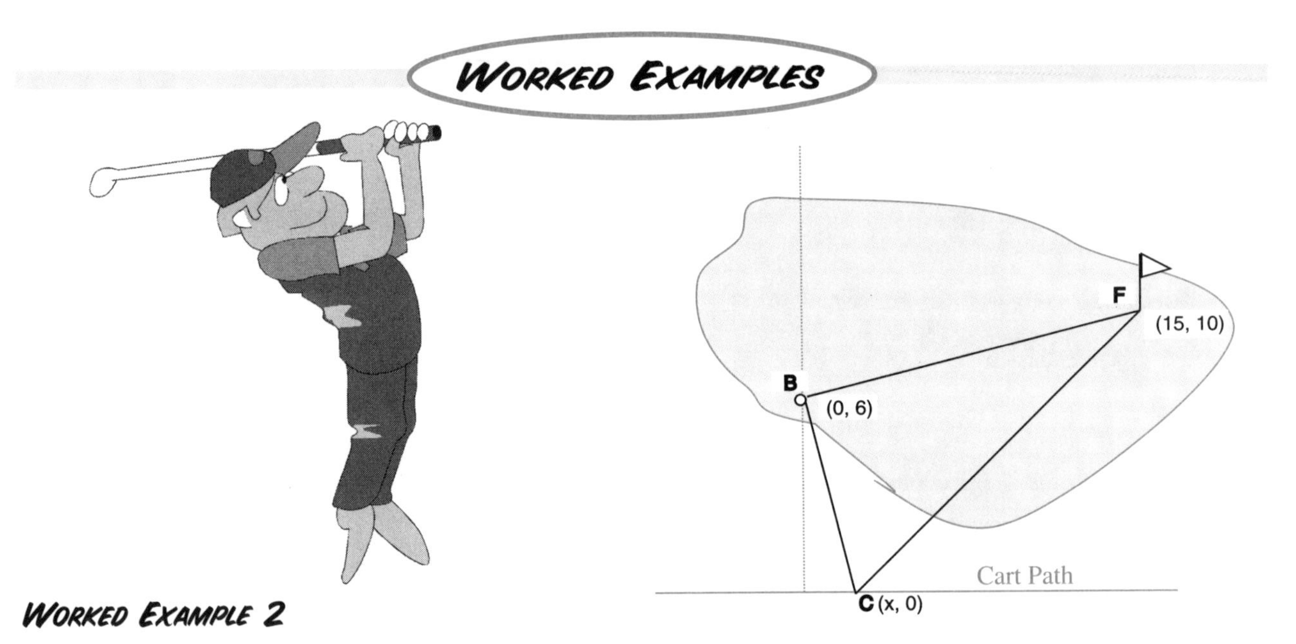

WORKED EXAMPLE 2

Golf carts are mandatory at the exclusive St. Patrick's Golf and Country Club. To speed up play, golfers are requested to park their carts on the cart path beside each green in such a way that the trip: cart → ball → flag → cart is a minimum distance. That is, golfers are required to park their carts in the position C(x, 0) for which the distance CB + BF + FC is a minimum.

Suppose the positions of the ball and flag are respectively (0, 6) and (15, 10) relative to the cart path as the x-axis, and the perpendicular line through the golf ball as the y-axis. Find the coordinates of C(x, 0) which will minimize CB + BF + FC.

SOLUTION

Since BF the distance from the ball to the flag is fixed, it is necessary only to minimize CB + FC.

$CB + FC = \sqrt{x^2+36} + \sqrt{(15-x)^2+10^2}$ ← applying the Pythagorean Theorem

That is, we must find a local minimum of the function $y = \sqrt{x^2+36} + \sqrt{(15-x)^2+10^2}$

In *Worked Example* 1 p. 34, we learned that when a function has a local extremum its first derivative is equal to zero. To find a local minimum of y, we calculate:

$$\frac{dy}{dx} = \frac{d}{dx}\left(\sqrt{x^2+36} + \sqrt{(15-x)^2+10^2}\right)$$

$$= \frac{d}{dx}\left(\sqrt{x^2+36}\right) + \frac{d}{dx}\left(\sqrt{(15-x)^2+10^2}\right)$$

$$= \frac{x}{\sqrt{x^2+36}} + \frac{x-15}{\sqrt{x^2-30x+325}}$$ ← from *Worked Example* 1

Setting $\frac{dy}{dx} = 0$ and solving yields $x = 45/8 = 5.625$ from which we calculate $y = \sqrt{481}$. That is, the distance is a minimum of 21.931…when the cart is at (5.625, 0).

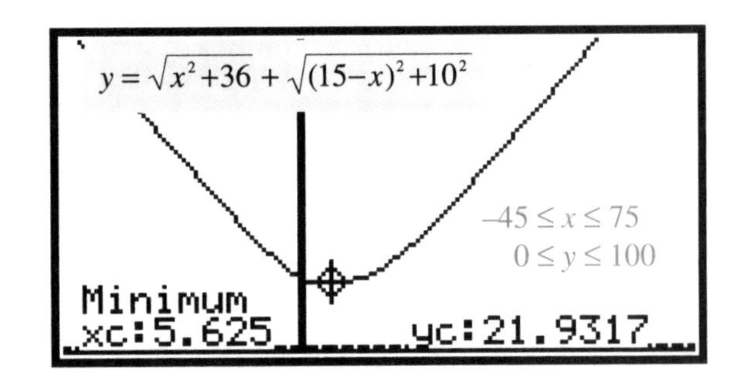

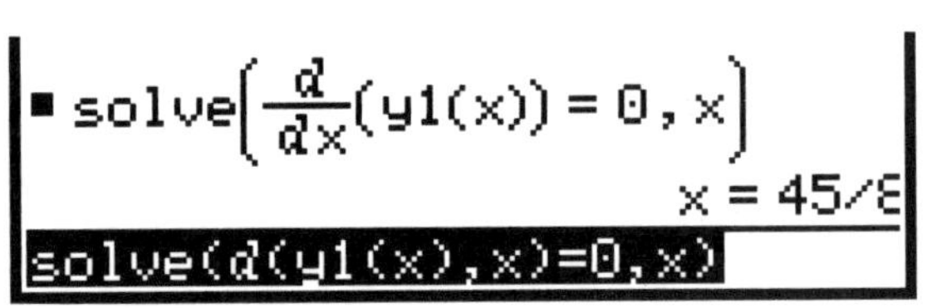

We can use the TI-89 to verify our solution. We define our function as $y1(x)$ and graph it as shown in the top display. The **Minimum** command on the F5 menu of the graph screen gives an approximation. For an exact solution in x, we enter the commands in the middle display. We then compute the corresponding y value as shown in the bottom display.

Exercises & Investigations

❶. For each function $y(x)$, state $\frac{dy}{dx}$.

a) $y = x^{\frac{1}{4}}$ b) $y = x^{-\frac{1}{2}}$ c) $y = -3x^{\frac{1}{3}}$

❷. Differentiate each function with respect to x.

a) $y = \sqrt[3]{x} + 3\sqrt[4]{x} - x^{-2}$ b) $y = \frac{1}{\sqrt{x}} + \sqrt{x}$

c) $y = \left(3+\sqrt{x}\right)^2$ d) $y = \left(\frac{1}{\sqrt{x}}+\sqrt{x}\right)^2$

❸. Can you apply the power rule directly to differentiate the expression $\sqrt{1+2x}$ with respect to x? Explain why or why not.

❹. Use the generalized power rule to differentiate each function with respect to x.

a) $y = \sqrt{5+3x}$ b) $y = \sqrt{6x^2-3x+1}$

c) $y = \sqrt[3]{5x^2-7x}$ d) $y = \frac{1}{\sqrt{6x^2-3x+1}}$

❺. Use the generalized power rule to differentiate each function with respect to x.

a) $y = (3x+5)^4$ b) $y = \sqrt{2x-7}$

c) $y = (2x-1)^3(3x+4)^2$ d) $y = (5+3x)^{-\frac{1}{2}}$

e) $y = \frac{(2x-1)^3}{3x+4}$ f) $y = \sqrt{\frac{3x^2+1}{5x-2}}$

❻. Calculate the values of x at which $\frac{dy}{dx} = 0$.

a) $y = 4x^3 - 11x^2 + 6x + 7$ b) $y = \sqrt{x^2+6x+3}$

c) $y = \frac{x^2}{\sqrt{4-x}}$ d) $y = \left(x^2+4\right)^2\left(2x^3-1\right)^3$

e) $y = (2x-1)\sqrt{3-x^2}$ f) $y = \sqrt{\frac{3x^2+1}{5x-2}}$

Note: To solve these equations on your TI-89, enter $y(x)$ on the Y= screen of your TI-89. Then enter the command shown below on the command line of your home screen.

```
solve(d(y1(x),x)=0,x)
```

❼. a) Graph each of the functions in exercise ❻ to check your answers.

b) Use the **Maximum** and **Minimum** commands on the F5 menu of your graph screen to determine where the local extrema of each function occur.

❽. Solve *Worked Example* 2 by marking the position F* of the image of the flag F under a reflection in the cart path. Then compare the distances BC + CF and BC + CF*, and slide C to minimize BC + CF*.

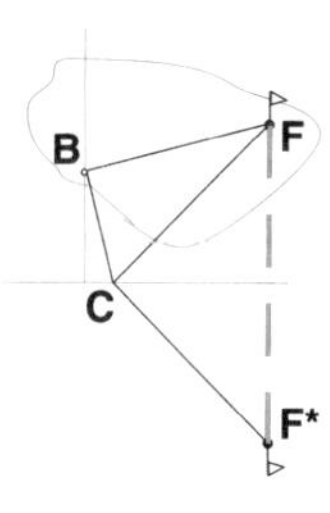

❾. A beam with rectangular cross section is to be cut from a cylindrical log of radius r. What dimensions of the rectangular cross section will yield the largest beam?

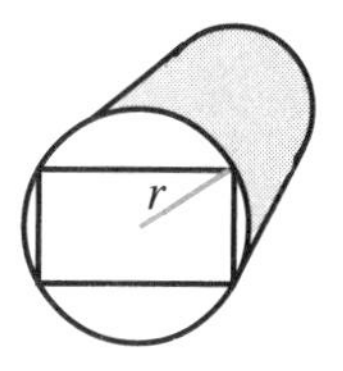

Express your answer as a function of r.

❿. Use the technique in *Worked Example* 1 to derive the **generalized power rule** where m is any rational number.

$$\frac{d}{dx}\left(f(x)^m\right) = mf(x)^{m-1}\frac{d}{dx}\left(f(x)\right)$$

⓫. Find the dimensions of a right circular cone of minimum volume that can circumscribe a sphere of radius 16 cm.

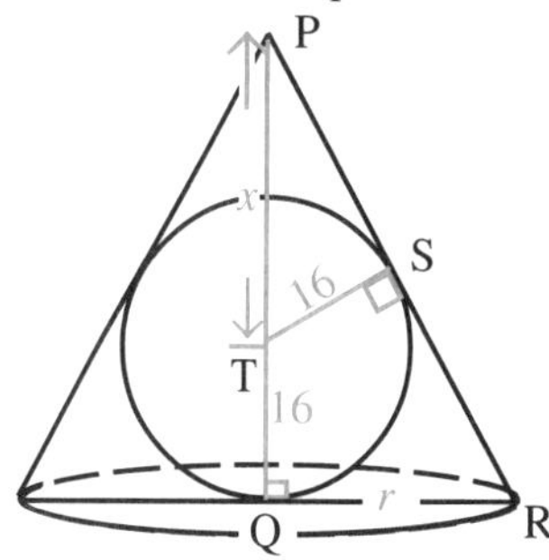

a) Write the volume V of the cone as a function of x and r.

b) Write the length of side PS in terms of x.

c) Explain how you know that ΔPST is similar to ΔPQR.

d) Use the similarity in part c) to find a ratio equal to 16/PS.

e) Solve for r in the proportion obtained in part d).

f) Substitute your expression for r into your expression for V so that V is expressed as a function of x only.

g) Minimize V by setting $dV/dx = 0$ and solving.

h) Graph V as a function of x and use the Minimum command on the F5 menu of your graph screen to check your answer.

HOW FAST DOES MONEY GROW WHEN COMPOUNDED INSTANTANEOUSLY?

In investigation ⓬ on page 24, you may have discovered that

$$\lim_{x\to\infty}\left(1+\frac{1}{x}\right)^x = e \quad \text{where } e \approx 2.71828\ldots$$

The number e like its companion π, is ubiquitous in mathematics. It serves as the base of the so-called exponential functions; i.e., functions of the form $y = e^{f(x)}$.

The exponential functions are accessed on your TI-89 by pressing: [◆] [e^x]. To graph $y = e^x$, press:

With the default window settings: $-10 \le x \le 10$; $-10 \le y \le 10$, we obtain the following display.

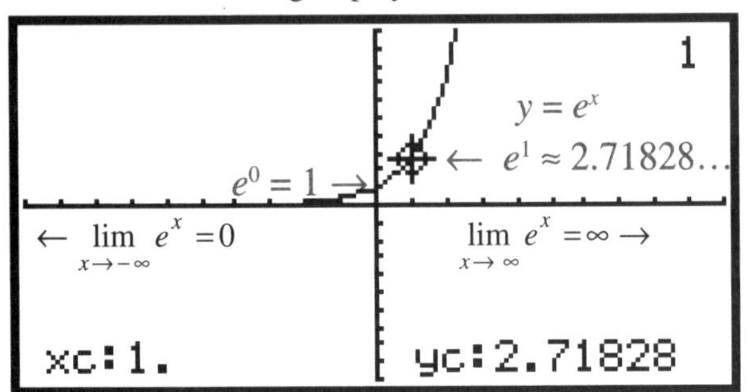

We observe in the display that:

$$e^0 = 1,\ e^1 \approx 2.71828\ldots, \quad \lim_{x\to-\infty} e^x = 0 \text{ and } \lim_{x\to\infty} e^x = \infty$$

A dollar invested at a nominal interest rate I per annum, compounded x times per year, grows to $\left(1+\frac{I}{x}\right)^x$ dollars in a year. As the number of compoundings is increased to instantaneous compounding, a dollar grows to

$$\lim_{x\to\infty}\left(1+\frac{I}{x}\right)^x = e^I \quad \text{dollars.}$$

x	y1
.1	1.1052
.11	1.1163
.12	1.1275
.13	1.1388
.14	1.1503

y1(x)=1.150273

The table shows the values of an investment of $1 at the end of one year for interest rates between 10% and 14% when compounding is instantaneous. We observe that a nominal rate of 14% is actually a rate of 15% when the compounding is instantaneous.

When quantities grow in proportion to their present value, the growth is said to be *exponential*. Consequently exponential functions are used to model population growth, radioactive decay, compound interest, rates of cooling, and other dynamical systems where rates of change are related to present values. Expressed mathematically, exponential functions are those that have their derivative at any point proportional to their value at that point.

The display shows the tangent line to the graph of $y = e^x$ at $x = 3$. Observe that the slope of the tangent is $e^3 \approx 20.0855\ldots$, the same as the value of e^x at $x = 3$. That is,

$$\left.\frac{dy}{dx}\right|_{x=3} = e^3$$

↑ slope at $x = 3$ ↑ $y(3)$

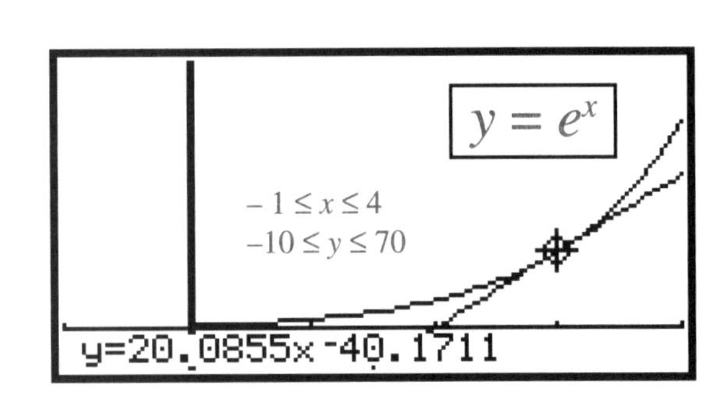

In what follows, we will prove that for all real x:

Exponential Rule $\boxed{\frac{d}{dx}(e^x) = e^x}$

To derive a formula for $\frac{d}{dx}(e^x)$, we apply Newton's binomial expansion for rational exponents.

$$\frac{d}{dx}(e^x) = \lim_{\Delta x \to 0} \frac{e^{x+\Delta x} - e^x}{\Delta x} = e^x \lim_{\Delta x \to 0} \frac{e^{\Delta x} - 1}{\Delta x}$$ ← using the definition of the derivative

$$= e^x \lim_{\Delta x \to 0} \left(\frac{\left(\lim_{h \to 0} (1+h)^{\frac{1}{h}} \right)^{\Delta x} - 1}{\Delta x} \right) = e^x \lim_{\Delta x \to 0} \left(\frac{\lim_{h \to 0} (1+h)^{\frac{\Delta x}{h}} - 1}{\Delta x} \right)$$ ← replacing $1/x$ by h in the defintion of e.

$$= e^x \lim_{\Delta x \to 0} \frac{\lim_{h \to 0} \left[1 + \frac{\Delta x}{h}(h) + \frac{\frac{\Delta x}{h}\left(\frac{\Delta x}{h} - 1\right)}{2!} h^2 + \frac{\left(\frac{\Delta x}{h}\right)\left(\frac{\Delta x}{h} - 1\right)\left(\frac{\Delta x}{h} - 2\right)}{3!} h^3 + \ldots \right] - 1}{\Delta x}$$ ← applying Newton's binomial expansion

$$= e^x \lim_{\Delta x \to 0} \frac{\lim_{h \to 0} \left[1 + \left(\frac{\Delta x}{h}\right) h \right] - 1}{\Delta x} = e^x$$ ← all powers of h above the first power approach 0 in the limit.

In investigation ⓬ on page 48, you will apply the method above to derive the general formula:

Generalized Exponential Rule

$$\frac{d}{dx}\left(e^{f(x)}\right) = e^{f(x)} \frac{d}{dx} f(x)$$

Worked Example

Differentiate with respect to x a) e^{3+x} b) e^{-x^2} c) $\frac{78.12}{6.3 + 102e^{-0.02817x}}$

Solution

a) $\frac{d}{dx} e^{3+x} = e^{3+x} \frac{d}{dx}(3 + x)$

To check this on your TI-89, press **F3** **ENTER** and enter the function as shown in the command line of the display.

```
■ d/dx(e^(3+x))          e^(x+3)
d(e^(3+x),x)
```

b) $\frac{d}{dx} e^{-x^2} = e^{-x^2} \frac{d}{dx}(-x^2) = -2xe^{-x^2}$

As above, we check this on the TI-89, obtaining this display.

```
■ d/dx(e^(-x^2))         -2·x·e^(-x^2)
d(e^(-x^2),x)
```

c) $\frac{d}{dx}\left(\frac{78.12}{6.3+102e^{-0.02817x}} \right) = \frac{d}{dx}(78.12)\left[6.3+102e^{-0.02817x}\right]^{-1}$

$$= -78.12(6.3 + 102e^{-0.02817x})^{-2} \frac{d}{dx}\left[6.3+102e^{-0.02817x}\right]$$

$$= -78.12(102)(-0.02817)\left[6.3+102e^{-0.02817x}\right]^{-2} e^{-0.02817x}$$

$$\approx \frac{224.46}{\left[6.3+102e^{-0.02817x}\right]^2 e^{0.02817x}}$$ ← In investigation ⓫, you will verify this answer on the TI-89.

EXERCISES & INVESTIGATIONS

❶. a) For what values of x is the function e^x defined?
b) For what values (if any) is the function e^x negative?
c) Does the exponential function e^x have any local extrema?
Give reasons for your answers.

❷. a) Graph the function $y = e^{-x}$. Explain how the graphs of $y = e^x$ and $y = e^{-x}$ are related.
b) Describe how the slopes of the tangents to the curves defined by $y = e^x$ and $y = e^{-x}$ are related.

❸. Graph the function $y = e^x$ in the window $-1 \le x \le 4; -1 \le y \le 70$
a) Evaluate e^x at these values.
(i) $x = 2$ (ii) $x = 3$ (iii) $x = 3.5$
b) Draw the tangent line at each value of x in part a). Check that the slope of the tangent in each case is equal to the value of e^x.

❹. Differentiate with respect to x.

a) $y=e^{3x^2+5x+1}$ b) $y=e^{-\frac{1}{x}}$

c) $y=e^{\sqrt{3x^2+5}}$ d) $y=\dfrac{1+e^x}{1-e^{-x}}$

❺. Evaluate the derivative of each of the functions in exercise ❹ at the point $x = 0.1$ and give the equation of the tangent line at that point.

❻. The derivative of a function $y = f(x)$ with respect to x is denoted $\frac{dy}{dx}$. The derivative of $\frac{dy}{dx}$ with respect to x is $\dfrac{d\left(\frac{dy}{dx}\right)}{dx}$ and it is denoted $\frac{d^2y}{dx^2}$. It is called the *second derivative of y with respect to x.*

Calculate $\frac{d^2y}{dx^2}$ of each of these functions.

a) $y = e^x$ b) $y = e^{-2x}$ c) $y = xe^x$

❼. Use the **solve(** command (F2 ENTER) on the Algebra menu of your TI-89 to find a numerical solution to the equation $e^x - e^{-x} = 1$.

❽. Find all the local extrema of the functions:

$$y = xe^{-x^2} \text{ and } y = x^2e^{-x}$$

a) by using the **Maximum** and **Minimum** commands on the F5 menu of the graph screen.
b) by using the **solve (** command on the Algebra menu to solve the equation $\frac{dy}{dx} = 0$.

❾. a) Find all the local extrema of the function defined by $y = x^3e^x$, using the F5 menu on your graph screen.
b) Use the **d(** command and solve the equation $\frac{dy}{dx}=0$, to check your answer in part a).

❿. a) Graph the function $y = \dfrac{x}{1-e^{-\frac{1}{x-2}}}$ in the window $-4 \le x \le 4; 1 \le y \le 2$.
b) Find all the local extrema for the function defined in part a) using the F5 menu of the graph screen.
c) Differentiate y with respect to x using the **d(** command.
d) Copy and paste the derivative you obtained in part c) into the command line of the home screen and solve $\frac{dy}{dx} = 0$ to check your answer to part b).

⓫. Use your TI-89 to verify the answer to part c) of the *Worked Example* on page 47.
(**Note:** $e^{0.02817} \approx 1.02857$)

⓬. Use the method on page 47 to prove the *generalized exponential rule*:

$$\frac{d}{dx}\left(e^{f(x)}\right) = e^{f(x)} \frac{d}{dx} f(x)$$

⓭. The *hyperbolic sine* and *hyperbolic cosine* are defined respectively by the equations:

$$\sinh(x) = \frac{e^x - e^{-x}}{2} \text{ and } \cosh(x) = \frac{e^x + e^{-x}}{2}.$$

a) Evaluate $\frac{d}{dx}(\sinh(x))$ and $\frac{d}{dx}(\cosh(x))$.
b) Express your answers in part a) in terms of $\sinh(x)$ and $\cosh(x)$.
c) Prove: $\cosh^2(x) - \sinh^2(x) = 1$
d) Graph $y = \sinh(x)$ and $y = \cosh(x)$ in the window $-4 \le x \le 4; -2 \le y \le 4$. Note: You can access the $\sinh(x)$ and $\cosh(x)$ functions on your TI-89 using the catalog key, CATALOG .
e) Find all the local extrema of $\sinh(x)$ and $\cosh(x)$.
f) Do the graphs of $\sinh(x)$ and $\cosh(x)$ intersect? To find out, select 5 (Intersection) on the F5 menu of the graph screen. What do you discover?
g) On the home screen, enter the commands shown in this display.

```
solve(sinh(x)=cosh(x),x)
```

What response does the display yield?
h) Verify the result in part g) by solving the equation $\sinh(x) = \cosh(x)$ without your calculator.

DERIVATIVES OF LOGARITHMIC FUNCTIONS

The exponential function $y = e^x$ expresses y as a function of x. To express x as a function of y, we define the *natural logarithm* (denoted ln and pronounced "lawn") to be the inverse of $y = e^x$. That is, $y = e^x \Rightarrow x = \ln y$. In other words, ln y is the power to which e must be raised to yield y. Expressed as a function of x:

ln x is the power to which e must be raised to yield x.

The natural logarithm function has the same properties as its common logarithm cousin.

Property 1	$\ln ab = \ln a + \ln b$	$a, b > 0$
Property 2	$\ln a/b = \ln a - \ln b$	$a, b > 0$
Property 3	$\ln a^b = b\ln a$	$a, b > 0$

Since $\ln x$ is inverse to e^x, then $x = e^{\ln x} = \ln e^x$. Furthermore, the graph of $\ln x$ can be obtained from the graph of e^x by a reflection in the line $y = x$.

Fechner's Law, enunciated in 1853 suggests an interesting relationship between the magnitude of a physical stimulus R and the intensity S of our sensation of it. In mathematical terms,

$S = k \ln R$, where k is a constant of proportionality.

This means that when a series of stimuli increase geometrically in intensity, we perceive them as increasing arithmetically. Consequently, a star that is perceived to be one magnitude brighter than another star is actually 2.5 *times* as bright.

We observe from the graph:

- $\ln x$ is defined for $x > 0$
- $\ln x < 0$ for $x < 1$
- $\ln x > 0$ for $x > 1$
- $\ln x = 0$ for $x = 1$
- $\lim_{x\to 0} \ln(x) = -\infty$
- $\lim_{x\to\infty} \ln(x) = \infty$

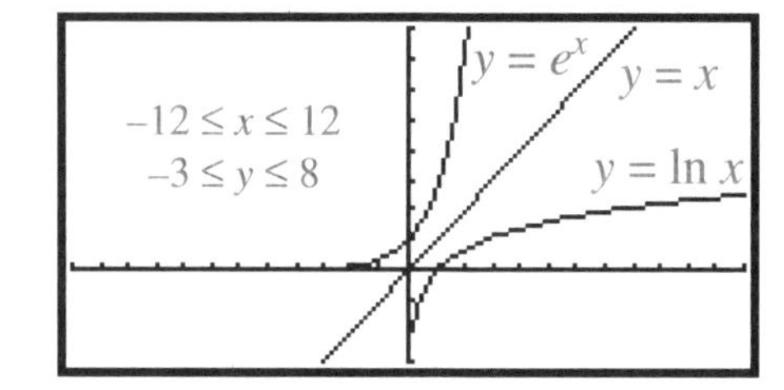

To obtain a formula for $\frac{d}{dx}(\ln(x))$, we use the above properties of the ln(x) function.

$$\frac{d}{dx}(\ln(x)) = \lim_{\Delta x\to 0} \frac{\ln(x+\Delta x)-\ln(x)}{\Delta x} = \lim_{\Delta x\to 0} \frac{\ln\left(\frac{x+\Delta x}{x}\right)}{\Delta x} \quad \text{by property 2}$$

$$= \lim_{\Delta x\to 0} \ln\left(\frac{x+\Delta x}{x}\right)^{\frac{1}{\Delta x}} \quad \text{by property 3}$$

$$= \frac{x}{x} \lim_{\Delta x\to 0} \ln\left(\frac{x+\Delta x}{x}\right)^{\frac{1}{\Delta x}} \quad \text{multiplying by } x / x$$

$$= \frac{1}{x} \lim_{\Delta x\to 0} \ln\left(\frac{x+\Delta x}{x}\right)^{\frac{x}{\Delta x}} \quad \text{by property 3}$$

$$= \frac{1}{x} \ln\left(\lim_{\Delta x\to 0}\left(\frac{x+\Delta x}{x}\right)^{\frac{1}{\Delta x}}\right) \quad \text{the log of a limit is the limit of a log (continuity)}$$

$$= \frac{1}{x}\ln e = \frac{1}{x} \quad \text{and so} \quad \boxed{\frac{d}{dx}\ln(x) = \frac{1}{x}}$$

Logarithmic Rule

In investigation 14 on page 52 you will apply the method above to prove the *generalized logarithm rule.*

Generalized Logarithmic Rule

$$\boxed{\frac{d}{dx}(\ln f(x)) = \frac{1}{f(x)}\frac{d}{dx}(f(x))}$$

WORKED EXAMPLES

The generalized exponent and logarithm formulas enable us to compute the derivatives of exponential and log functions with base e; but how do we differentiate functions of the form $y = a^{f(x)}$ and $y = \log_a f(x)$ where $a > 0$ and $a \neq e$? To differentiate $y = a^{f(x)}$ we convert the base to e by using the identity $a = e^{\ln a}$ to substitute for a, as shown below.

$$\frac{d}{dx}\left(a^{f(x)}\right) = \frac{d}{dx}\left(e^{\ln a}\right)^{f(x)} = \frac{d}{dx}e^{(\ln a)f(x)} = e^{(\ln a)f(x)}\frac{d}{dx}\left((\ln a)f(x)\right) = a^{f(x)}\ln a\frac{d}{dx}(f(x))$$

That is, $\boxed{\frac{d}{dx}\left(a^{f(x)}\right) = a^{f(x)}\ln a\frac{d}{dx}(f(x))}$ ← We observe that when an exponential function is expressed in base a, we merely differentiate it as though it were in base e and then adjust by multiplying by $\ln a$.

In a similar way, we can work from the formula for the derivative of a log function to base e and generalize it to apply to an arbitrary base a by using the identity $a = e^{\ln a}$.

To evaluate $\frac{d}{dx}\left(\log_a f(x)\right)$, we express $\log_a f(x)$ in terms of $\ln f(x)$ by writing,

$$f(x) = a^{\log_a f(x)} = \left(e^{\ln a}\right)^{\log_a f(x)} = e^{\ln a \log_a f(x)}$$

so, $\ln f(x) = (\ln a)\log_a f(x)$, i.e. $\log_a f(x) = \frac{\ln f(x)}{\ln a}$.

$$\frac{d}{dx}\left(\log_a f(x)\right) = \frac{d}{dx}\left(\frac{1}{\ln a}\ln f(x)\right) = \frac{1}{\ln a}\frac{d}{dx}\left(\ln f(x)\right) = \frac{1}{\ln a}\frac{1}{f(x)}\frac{d}{dx}(f(x))$$

That is, $\boxed{\frac{d}{dx}\left(\log_a f(x)\right) = \frac{1}{(\ln a)f(x)}\frac{d}{dx}(f(x))}$ ← We observe that when a logarithmic function is expressed in base a, we merely differentiate it as though it were in base e and then adjust by dividing by $\ln a$.

WORKED EXAMPLE 1

Differentiate with respect to x.

a) $3^{-\frac{x^2}{1+x^2}}$ b) x^x

DIFFERENTIATE ON YOUR TI-89 AND SHOW THAT EACH ANSWER ON YOUR DISPLAY IS EQUIVALENT TO THE ONE GIVEN HERE.

SOLUTION

a) $$\frac{d}{dx}3^{-\frac{x^2}{1+x^2}} = 3^{-\frac{x^2}{1+x^2}}\ln 3\frac{d}{dx}\left(\frac{-x^2}{1+x^2}\right) = 3^{-\frac{x^2}{1+x^2}}\ln 3\frac{d}{dx}\left(-x^2\left(1+x^2\right)^{-1}\right)$$

$$= 3^{-\frac{x^2}{1+x^2}}\ln 3\left(-x^2\frac{d}{dx}\left(1+x^2\right)^{-1} + \left(1+x^2\right)^{-1}\frac{d}{dx}\left(-x^2\right)\right)$$

$$= 3^{-\frac{x^2}{1+x^2}}\ln 3\left(x^2\frac{2x}{\left(1+x^2\right)^2} + \frac{-2x}{\left(1+x^2\right)}\right) = -3^{-\frac{x^2}{1+x^2}}(\ln 3)\left(\frac{2x}{\left(1+x^2\right)^2}\right)$$

b) $$\frac{d}{dx}x^x = \frac{d}{dx}\left(e^{\ln x}\right)^x = \frac{d}{dx}\left(e^{x\ln x}\right) = e^{x\ln x}\frac{d}{dx}(x\ln x) = e^{x\ln x}(1+\ln x) = x^x(1+\ln x)$$

OBSERVE HOW WE DIFFERENTIATE x^x BY CONVERTING IT TO $e^{x\ln x}$.

Worked Example 2

Differentiate with respect to x.

a) $\ln(x^2+1)$ b) $\ln\sqrt{3x^3+2x^2+1}$ c) $\log_2\left(\dfrac{3x^2-1}{5x^3+x}\right)$

Solution

a) $\dfrac{d}{dx}\ln(x^2+1) = \dfrac{1}{x^2+1}\dfrac{d}{dx}(x^2+1) = \dfrac{2x}{x^2+1}$

b) $\dfrac{d}{dx}\left(\ln\sqrt{3x^3+2x^2+1}\right) = \dfrac{1}{\sqrt{3x^3+2x^2+1}}\dfrac{d}{dx}\left(\sqrt{3x^3+2x^2+1}\right) = \dfrac{(9x^2+4x)}{2\left(3x^3+2x^2+1\right)}$

c) $\dfrac{d}{dx}\left(\log_2\left(\dfrac{3x^2-1}{5x^3+x}\right)\right) = \dfrac{d}{dx}\left(\log_2\left(3x^2-1\right) - \log_2\left(5x^2+1\right) - \log_2 x\right) = \dfrac{1}{\ln 2}\left(\dfrac{3x^2+1}{x\left(3x^2-1\right)} - \dfrac{10x}{5x^2+1}\right)$

Worked Example 2c) shows how we can use the properties of the log function to simplify the differentiation process when products or quotients are involved.

If $y = f(x)$, then $\dfrac{d}{dx}(\ln y) = \dfrac{1}{y}\dfrac{dy}{dx}$, so $\boxed{\dfrac{dy}{dx} = y\dfrac{d}{dx}(\ln y)}$

Therefore, if y is a product or quotient, we can compute $\dfrac{dy}{dx}$ by merely finding $\dfrac{d}{dx}(\ln y)$ and multiplying the result by y. This technique is called *logarithmic differentiation.* The next example shows how we apply logarithmic differentiation.

Worked Example 3

Use logarithmic differentiation to calculate the derivative with respect to x.

a) $y = \sqrt{\dfrac{x^3-1}{x^2+1}}$ b) $y = \dfrac{x^3\sqrt{5-3x}}{(3x+1)^2}$

Solution

a)
$$\frac{dy}{dx} = y\frac{d}{dx}(\ln y)$$
$$= y\frac{d}{dx}\left(\ln\sqrt{x^3-1} - \ln\sqrt{x^2+1}\right)$$
$$= \sqrt{\frac{x^3-1}{x^2+1}}\left(\frac{1}{2}\right)\left(\frac{3x^2}{x^3-1} - \frac{2x}{x^2+1}\right)$$
$$= \frac{x^4+3x^2+2x}{2(x^3-1)^{\frac{1}{2}}(x^2+1)^{\frac{3}{2}}}$$

b)
$$\frac{dy}{dx} = y\frac{d}{dx}(\ln y)$$
$$= y\frac{d}{dx}\left(\ln x^3\sqrt{5-3x} - \ln(3x+1)^2\right)$$
$$= \frac{x^3\sqrt{5-3x}}{(3x+1)^2}\left(\frac{d}{dx}(\ln x^3) + \frac{d}{dx}\left(\ln\sqrt{5-3x}\right) - \frac{d}{dx}\left(\ln(3x+1)^2\right)\right)$$
$$= \frac{x^3\sqrt{5-3x}}{(3x+1)^2}\left(\frac{3}{x} - \frac{3}{2(5-3x)} - \frac{6}{(3x+1)}\right) = \frac{x^3\sqrt{5-3x}}{(3x+1)^2}\left(\frac{27x^2-9x-30}{18x^3-24x^2-10x}\right)$$

Exercises & Investigations

1. Explain why each identity is true.
a) $e^{\ln x} = x$ b) $\ln e^x = x$

2. a) For what values of x does $\ln x$ exist?
b) For what values of x is $\ln x < 1$?
c) For what values of x is $\ln x > 1$?

3. For what values of x is the slope of the graph of $y = \ln x$:
a) < 1?   b) > 1?   c) $= 1$?   d) $= 2$?

4. Given $\ln x = y$, express $\log_a y$ in terms of x and a.

5. Differentiate with respect to x.

a) 2^{-x^2} b) x^{x^2} c) $(2x+1)^x$

6. Differentiate $3^{-\frac{x^2}{1+x^2}}$ with respect to x on your TI-89 using the **d (** command. Compare your answer with the answer given in *Worked Example* 1 on page 50 and verify that both answers are equivalent.

7. Differentiate with respect to x.

a) $\ln(3x^2+2)$ b) $\ln\sqrt{x^2-3}$ c) $\log_2\left(\frac{1}{\sqrt{x^2+1}}\right)$

8. a) Explain the meaning of *logarithmic differentiation.*
b) Use logarithmic differentiation to differentiate each of the following with respect to x.

i) $y = \sqrt{\frac{x^2+1}{x^2+4}}$ ii) $y = \sqrt[3]{\frac{x}{3x+5}}$ iii) $y = \sqrt{\frac{x^2+2x-1}{x^2+5}}$

9. Differentiate with respect to x.

a) $y = \frac{1}{1+\ln x}$

b) $y = \frac{\ln x}{x}$

c) $y = \ln\sqrt{\frac{(1+x)(1+2x)(1+3x)}{(1-x)(1-2x)(1-3x)}}$

10. a) Use logarithmic differentiation to calculate $\frac{dy}{dx}$ where $y = \sqrt{x(x^2-3)(x^3+2x+1)}$
b) Find the values of x at which y has local extrema.

11. Differentiate with respect to x.

a) $\log_{10}(3x^2+2)$ b) $\log_{10}\sqrt{x^2-3}$

c) $\log_{10}\left(\frac{3x^2+2}{\sqrt{x^2-3}}\right)$ d) $\log_{10}\left(\frac{\sqrt{x^2-3}}{3x^2+2}\right)$

12. Use logarithmic differentiation to prove each of the following.

a) $$\frac{d[f(x)g(x)]}{dx} = f(x)\frac{dg(x)}{dx} + g(x)\frac{df(x)}{dx}$$

b) $$\frac{d\left[\frac{f(x)}{g(x)}\right]}{dx} = \frac{g(x)\frac{df(x)}{dx} - f(x)\frac{dg(x)}{dx}}{g(x)^2}$$

13. Use logarithmic differentiation to prove the power rule.

$$\frac{d}{dx}\left(x^m\right) = m\,x^{m-1}$$

14. Follow the method on page 49 to prove the generalized logarithm rule.

$$\frac{d}{dx}(\ln f(x)) = \frac{1}{f(x)}\frac{d}{dx}(f(x))$$

How Many Prime Numbers up to N?

In 1896, Hadamard and de la Vallée Poussin proved the following theorem known as the *Prime Number Theorem.*

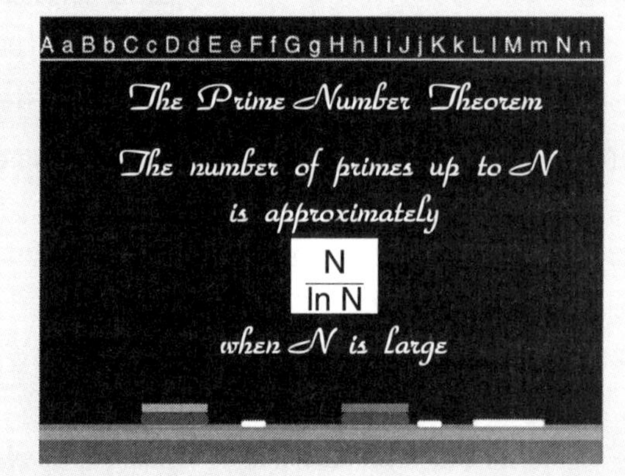

a) Graph the function $p(x) = \frac{x}{\ln x}$ in the window
$0 \le x \le 110;\ 0 \le y \le 30$

b) Use the **value** command on the F5 menu to evaluate $p(100)$. Compare with $\pi(100) = 25$, the actual number of primes up to 100.

c) The Locher-Ernst formula for approximating $\pi(x)$ is

$$P(x) = \frac{x}{\frac{1}{3}+\frac{1}{4}+\frac{1}{5}+\dots\ \frac{1}{x}}$$

d) Graph $P(x)$ and $p(x)$ in the same window as in part a). Compare $p(100)$ and $P(100)$ with the true value 25.

e) Repeat part d) to compare $p(1000)$ and $P(1000)$ with the true value 168. (The graphing takes a few minutes.)

f) Calculate $\lim_{x\to\infty} p'(x)$. What does this mean?

DERIVATIVES OF TRIGONOMETRIC FUNCTIONS

Many of Nature's processes are periodic. The seasons, the tides, life cycles of plants and animals, and even your heartbeat. In *Exploration* 16 of *Advanced Algebra with the* TI-89, (see page 96 of this book) we studied the function

$$y = \frac{35}{3} + \frac{7\sin\left(\frac{72(x-80)}{73}\right)}{3}$$

which expresses the number of hours of daylight, y which New Orleans receives on the xth day of the year. (Here we set $x = 1$ on January 1 and $x = 365$ on December 31, and the argument of the sine function is in degrees.) In that Exploration, we found, by tracing along the graph, the days of the year with minimum and with maximum hours of sunlight.

A sunrise in New Orleans

Now armed with calculus, we can find extrema by calculating $\frac{dy}{dx}$ and solving $\frac{dy}{dx} = 0$. We can then graph the function to determine whether we have a local extremum and whether it's a maximum or a minimum. (Alternatively, we can calculate $\frac{d^2y}{dx^2}$ and apply the second derivative test.) This procedure requires that we develop a formula for the derivative of the sine function.

From the definition of the sine function *in radians,* we derive a formula for the derivative of the sine function as follows.

$$\frac{d}{dx}(\sin x) = \lim_{\Delta x \to 0} \frac{\sin(x + \Delta x) - \sin(x)}{\Delta x} \quad \leftarrow \text{definition of a derivative}$$

$$= \lim_{\Delta x \to 0} \frac{2\cos\left(\frac{2x + \Delta x}{2}\right)\sin\left(\frac{\Delta x}{2}\right)}{\Delta x} \quad \leftarrow \boxed{\sin A - \sin B = 2\cos\left(\frac{A+B}{2}\right)\sin\left(\frac{A-B}{2}\right)}$$

$$= \cos(x) \lim_{\Delta x \to 0} \frac{2\sin\left(\frac{\Delta x}{2}\right)}{\Delta x}$$

$$= \cos(x) \lim_{\Delta x \to 0} \frac{\sin\left(\frac{\Delta x}{2}\right)}{\left(\frac{\Delta x}{2}\right)} \quad \leftarrow$$

Recall

$$\lim_{\theta \to 0} \frac{\sin\theta}{\theta} = 1$$

$$= \cos x$$

That is, $\boxed{\frac{d}{dx}(\sin x) = \cos x}$

In investigation 11 on page 56, you will use the procedure above to prove the companion formula. $\rightarrow$ $\boxed{\frac{d}{dx}(\cos x) = -\sin x}$

In investigation 12 you will generalize the process above to derive this formula for the derivative of the generalized sine function.

$\boxed{\frac{d}{dx}(\sin f(x)) = \cos(f(x))\,\frac{d}{dx}(f(x))}$ Similarly, $\boxed{\frac{d}{dx}(\cos f(x)) = -\sin(f(x))\,\frac{d}{dx}(f(x))}$

WORKED EXAMPLE

The formula $y = \frac{35}{3} + \frac{7\sin\left(\frac{72(x-80)}{73}\right)}{3}$ expresses the number of hours of daylight y which New Orleans has on the xth day of the year (where counting starts on January 1, and the argument of the sine function is in degrees). Graph y as a function of x.

Use your graph to determine:

a) The days on which New Orleans receives the fewest (most) hours of daylight.
b) The days on which the number of hours of sunlight is increasing (decreasing) most rapidly.

SOLUTION

To avoid awkward constants when we differentiate the sine function, we begin by modifying the formula for y so that the argument of the sine function is expressed in radians. To convert from degrees to radians, we multiply the argument of the sine function by $\pi/180$ to obtain:

$$y = \frac{35 + 7\sin\left(\frac{2\pi(x-80)}{365}\right)}{3}$$

To graph y as a function of x on the TI-89, we must first press: MODE ▼ ▼ ▼ and select RADIAN mode. Then press ENTER *twice*. Then we set the window variables to $0 \le x \le 730$ (to get two complete cycles) and $0 \le y \le 18$, because the number of hours of daylight will lie in this range. We obtain the graph shown in the display.

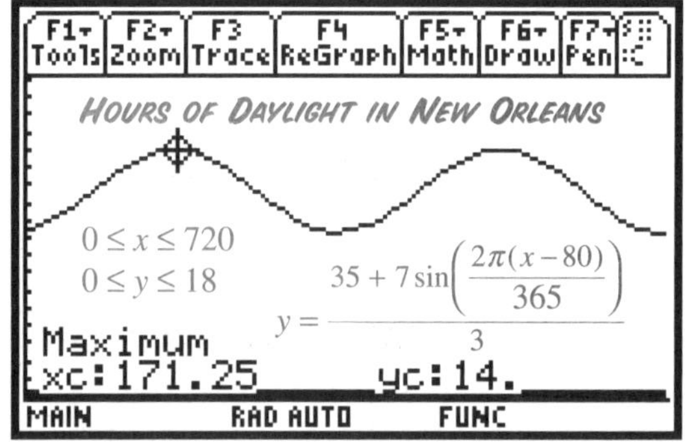

a) To find the local extrema, we can access in turn the commands, Maximum and Minimum from the F5 menu of the graph screen. Proceeding as in *Worked Example* 1 p. 34, we obtain a local minimum at (353.75, 9.33…) and a local maximum at (171.25, 14). That is, on the 354th day of the year; i.e., December 20, the number of hours of daylight reaches a minimum of 9 hours and 20 minutes, and on the 172nd day of the year; i.e., June 21, the number of hours of daylight reaches a maximum of 14 hours.

Alternatively, we compute:

$$\frac{dy}{dx} = \frac{7}{3}\left(\frac{2\pi}{365}\right)\cos\left(\frac{2\pi(x-80)}{365}\right)$$

Setting $\frac{dy}{dx} = 0$ yields $\cos\left(\frac{2\pi(x-80)}{365}\right) = 0$,

so $\frac{2\pi(x-80)}{365} = \frac{\pi}{2}$ or $\frac{3\pi}{2}$

Solving for x yields, $x = 171.25$ or $x = 353.75$ as found above.

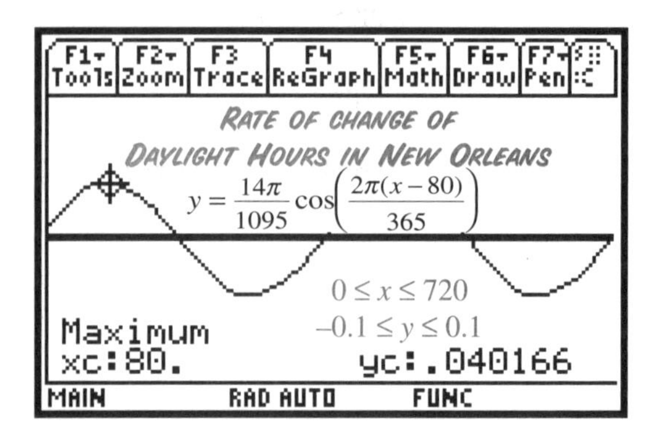

SOLUTION (cont'd)

b) The rate of change of daylight hours is given by $\frac{dy}{dx} = \frac{14\pi}{1095}\cos\left(\frac{2\pi(x-80)}{365}\right)$ (as found in part a).

To find the days when this change is greatest, we could graph this function and proceed as in part a) (see bottom display p. 54).

Alternatively, we can take the derivative of $\frac{dy}{dx}$ to obtain the *second derivative*

$$\frac{d^2y}{dx^2} = \frac{d\left(\frac{14\pi}{1095}\cos\left(\frac{2\pi(x-80)}{365}\right)\right)}{dx} = -\frac{28\pi^2}{(365)(1095)}\sin\left(\frac{2\pi(x-80)}{365}\right)$$

The rate of change of daylight hours has its extrema when the *derivative* of $\frac{dy}{dx}$ is 0, i.e. when $\frac{d^2y}{dx^2} = 0$.

This occurs when $\frac{2\pi(x-80)}{365} = 0$ or π i.e., when $x = 80$ or 262.5.

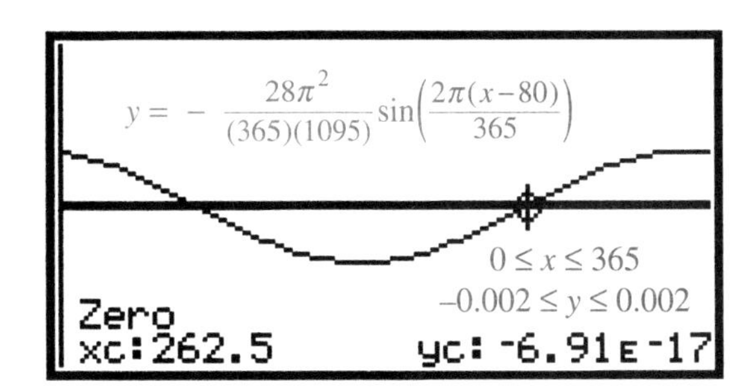

To find the rate of change of daylight hours at these extrema, we substitute these values of x into the expression for $\frac{dy}{dx}$ to obtain a maximum rate of increase of 0.04 hours/day (i.e., 2.4 min/day) on the 80th day (March 21) and a maximum rate of decrease of 0.04 hours/day on the 263rd day (September 20).

In the *Worked Example*, we calculated $\frac{dy}{dx}$ and $\frac{d^2y}{dx^2}$ for a given function y. Observe that in the graph of $\frac{dy}{dx}$, that $\frac{dy}{dx} = 0$ when y attains its local extrema. Furthermore, we see that $\frac{d^2y}{dx^2} < 0$ around a local maximum and $\frac{d^2y}{dx^2} > 0$ around a local minimum. The reason for this can be seen by observing the split-screen displays shown below. As y approaches a maximum, at $x = 171.25$ the slope $\frac{dy}{dx}$ decreases to zero and then continues to decrease through zero into negative values; i.e., $\frac{dy}{dx}$ changes its sign from positive to negative, indicating that the *rate of change of slope* near a local maximum for y is negative; i.e., $\frac{d^2y}{dx^2} < 0$. A similar argument explains why $\frac{d^2y}{dx^2} > 0$ at a local minimum. Therefore, to determine whether a local extremum is a maximum or a minimum, we can take the second derivative and apply the following test.

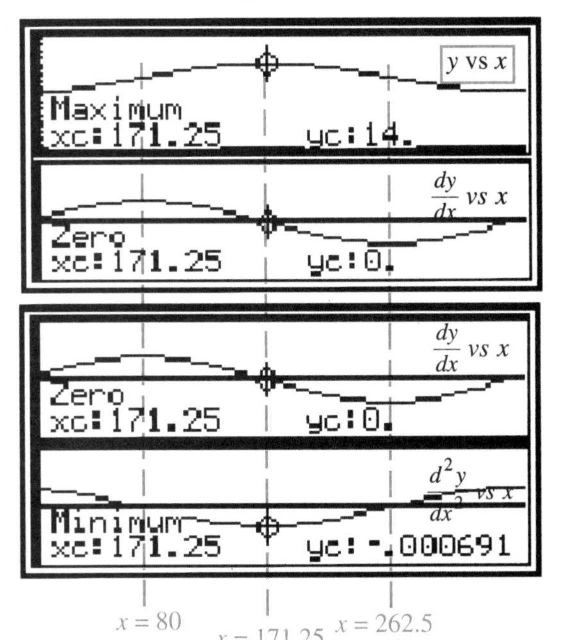

condition for local minimum	condition for local maximum
$\frac{dy}{dx} = 0$ and $\frac{d^2y}{dx^2} > 0$	$\frac{dy}{dx} = 0$ and $\frac{d^2y}{dx^2} < 0$

Definition If $\frac{d^2y}{dx^2} = 0$ for some value of x, we say y has a *point of inflection* at that value of x.

Note: To obtain a split screen, press: MODE F2

Press 2nd ⟨↔⟩ to move between screens.

EXERCISES & INVESTIGATIONS

1. Suppose $y = f(x)$ is a differentiable function.

a) Explain how the sign of $\frac{dy}{dx}$ changes as x increases at
(i) a local maximum of $f(x)$.
(ii) a local minimum of $f(x)$.

b) Explain how the sign of $\frac{d^2y}{dx^2}$ changes as x increases from a local minimum to a local maximum of $f(x)$.

c) Explain how to use $\frac{d^2y}{dx^2}$ to determine whether a local extremum of $f(x)$ is a local maximum or local minimum

d) Describe and define a *point of inflection.*

2. Explain how you could have used your knowledge of the sine and cosine functions to determine the extrema in the *Worked Example* directly; i.e., without graphing or computing $\frac{dy}{dx}$.

3. Differentiate with respect to x.

a) $\frac{d}{dx}\left(\sin^2 x\right)$ b) $\frac{d}{dx}\left(\sin(3x+5)\right)$ c) $\frac{d}{dx}\left(\cos(5x+1)\right)$

4. Differentiate with respect to x.

a) $\frac{d}{dx}(\sec x)$ b) $\frac{d}{dx}(\tan x)$ c) $\frac{d}{dx}\left(\csc(3x+1)\right)$

5. Calculate $\frac{dy}{dx}$ for $y = 3\sin(5x+6)$ where the argument is in degrees.

6. The horizontal displacement x of a simple pendulum from rest is given by the equation

$x = 4\cos\left(\frac{2\pi t}{3}\right)$ where t is elapsed time in seconds.

a) Write expressions for the velocity and acceleration of the pendulum as a function of time. (Recall that acceleration is the rate of change of velocity.)
b) Determine the times at which the pendulum reaches maximum and minimum displacement from the rest position.
c) At what times does the pendulum reach maximum velocity?
d) At what times does the pendulum reach maximum acceleration?

7. Use the d (command on your TI-89 to calculate $\frac{dy}{dx}$ for

$$y = \frac{35 + 7\sin\left(\frac{2\pi(x-80)}{365}\right)}{3}$$

Compare your answer with the answer in the *Worked Example* on p. 54. Prove both answers are equivalent.

8. Prove that $\sin^2 x + \cos^2 x = 1$, by differentiating the function $y = \sin^2 x + \cos^2 x$ and calculating $\frac{dy}{dx}$.

9. Calculate $\frac{d^2y}{dx^2}$ for each function.

a) $y = e^{\sin x}$ b) $y = xe^{\sin x}$ c) $\ln \cos x$

Use your TI-89 to check your answers.

10. Determine the points of inflection of the function

$$y = \frac{35 + 7\sin\left(\frac{2\pi(x-80)}{365}\right)}{3}$$

11. Follow the procedure on page 53 to derive the formula

$$\frac{d}{dx}(\cos x) = -\sin x$$

12. Generalize the procedure on page 53 to derive the formula $\frac{d}{dx}\left(\sin f(x)\right) = \cos(f(x))\,\frac{d}{dx}\left(f(x)\right)$

13. L'Hôpital's Rule

Given a function $y = f(x)$ such that $f(0) = 0$

a) Write an expression for $|PQ|$ in terms of $f'(0)$ and Δx, if OQ is tangent to $y = f(x)$ at $x = 0$.

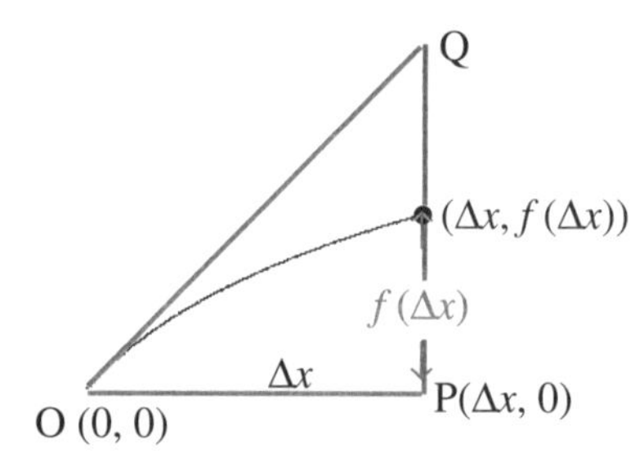

b) Explain why $\lim_{\Delta x \to 0} f(\Delta x) = \lim_{\Delta x \to 0} |PQ|$

c) Suppose $g(x)$ is also differentiable with $g(0) = 0$. Use your estimate in part a) to express

$$\frac{f(\Delta x)}{g(\Delta x)}$$

in terms of $f'(0)$ and $g'(0)$.

d) Evaluate the limit of the quotient in part c) as $\Delta x \to 0$. This result is known as l'Hôpital's Rule named by its discoverer *Johann Bernoulli* in honor of his patron, the *Marquis de l'Hôpital* (1661-1704).

e) Use l'Hôpital's Rule to evaluate these limits.

i) $\lim_{x\to 0} \frac{\sin x}{x}$ ii) $\lim_{x\to 1} \frac{\ln x^2}{\sin \pi x}$ iii) $\lim_{x\to 0}\left(\frac{1}{x} - \frac{1}{\sin x}\right)$

THE CHAIN RULE & RELATED RATES

The table below summarizes the formulas we have derived for differentiating simple and generalized functions.

	SIMPLE FORM	GENERALIZED FORM
POWER RULE	$\frac{d}{dx}(x^m) = mx^{m-1}$	$\frac{d}{dx}(f(x)^m) = mf(x)^{m-1}\frac{d}{dx}(f(x))$
EXPONENT RULE	$\frac{d}{dx}(e^x) = e^x$	$\frac{d}{dx}(e^{f(x)}) = e^{f(x)}\frac{d}{dx}f(x)$
LOGARITHM RULE	$\frac{d}{dx}\ln(x) = \frac{1}{x}$	$\frac{d}{dx}(\ln f(x)) = \frac{1}{f(x)}\frac{d}{dx}(f(x))$
SINE RULE	$\frac{d}{dx}(\sin x) = \cos x$	$\frac{d}{dx}(\sin f(x)) = \cos(f(x))\frac{d}{dx}(f(x))$
COSINE RULE	$\frac{d}{dx}(\cos x) = -\sin x$	$\frac{d}{dx}(\cos f(x)) = -\sin(f(x))\frac{d}{dx}(f(x))$

The generalized form of the formulas above are examples of the *chain rule,* which is stated formally below.

Theorem: The Chain Rule

If $f(x)$ and $g(x)$ are differentiable functions, then the derivative of the composite function $g(f(x))$ is given by

$$\frac{d}{dx}(g(f(x))) = \frac{dg(f(x))}{df(x)} \bullet \frac{df(x)}{dx}$$

Proof

$$\frac{d}{dx}(g(f(x))) = \lim_{\Delta x \to 0}\frac{g(f(x+\Delta x)) - g(f(x))}{\Delta x}$$ ← by definition of the derivative

$$= \lim_{\Delta x \to 0}\frac{g(f(x+\Delta x)) - g(f(x))}{f(x+\Delta x) - f(x)} \bullet \frac{f(x+\Delta x) - f(x)}{\Delta x}$$ ← Assuming $f(x+\Delta x) - f(x) \neq 0$, for $\Delta x > 0$, we can multiply and divide by this quantity.

$$= \lim_{\Delta x \to 0}\frac{g(f(x+\Delta x)) - g(f(x))}{f(x+\Delta x) - f(x)} \bullet \lim_{\Delta x \to 0}\frac{f(x+\Delta x) - f(x)}{\Delta x}$$ ← For continuous functions, the limit of a product is equal to the product of the limits.

$$= \frac{dg(f(x))}{df(x)} \bullet \frac{df(x)}{dx}$$

Note: The symbol $df(x)$ which appears in the numerator and denominator links the successive factors in the product like the links of a chain and hence the theorem is called the *chain rule.* This "chain" aspect is particularly evident when the rule is extended to compositions of more than two functions; e.g.,

$$\frac{d}{dx}h(g(f(x))) = \frac{dh(g(f(x)))}{dg(f(x))}\frac{dg(f(x))}{df(x)} \bullet \frac{df(x)}{dx}$$

WORKED EXAMPLES

WORKED EXAMPLE 1

The volume of a spherical hot air balloon is increasing at a rate of 3.5 m^3 per minute when its radius is $2m$. At what rate is its radius increasing?

Filling a balloon with hot air.

SOLUTION

If R denotes the radius of the balloon, and V its volume, then

$$V = \frac{4}{3}\pi R^3 \quad \leftarrow \text{because the balloon is spherical}$$

If we differentiate both sides of this equation with respect to the time t, we obtain

$$\frac{dV}{dt} = \frac{dV}{dR} \bullet \frac{dR}{dt} \quad \leftarrow \text{applying the chain rule}$$

$$= 4\pi R^2 \frac{dR}{dt} \quad ①$$

Equation 1 links two rates, the rate of change of the volume of air in the balloon dV/dt, and the rate of change of the radius, dR/dt.

We are given that $dV/dt = 3.5\ m^3/\text{min}$. Substituting this value into equation ① and solving for dR/dt yields

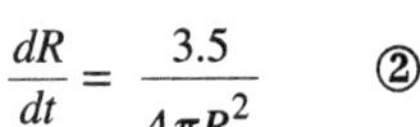

$$\frac{dR}{dt} = \frac{3.5}{4\pi R^2} \quad ②$$

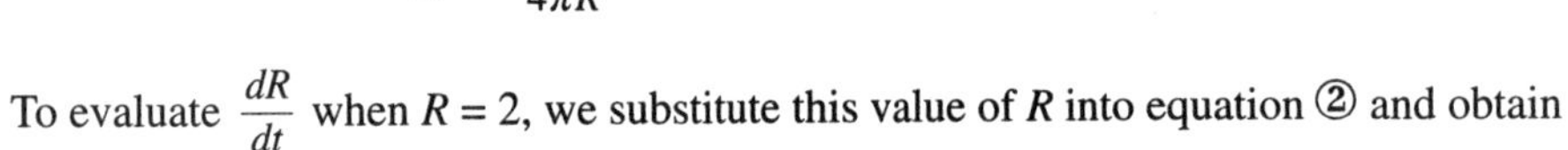

To evaluate $\frac{dR}{dt}$ when $R = 2$, we substitute this value of R into equation ② and obtain

$$\frac{dR}{dt} = \frac{3.5}{4\pi(2)^2} \approx 0.06963\ m/\text{min}$$

That is, the radius of the balloon is increasing at about 0.07 m/min or 7 cm/min when its radius is 2 m.

WORKED EXAMPLE 2

A kite 40 m above the ground is drifting horizontally at a speed of 2 m/s (meters per second). At what rate is the angle of inclination of the string changing when the string is 60 m long?

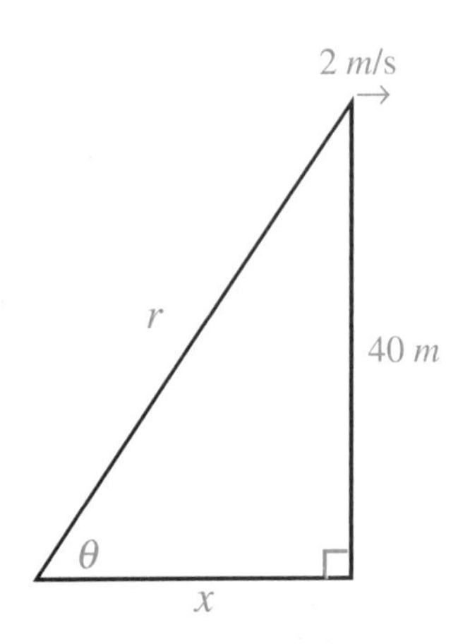

SOLUTION

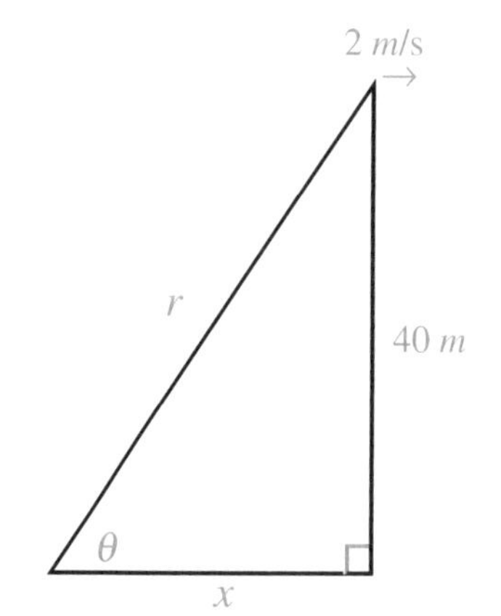

Let θ denote the angle of inclination at time t and let x denote the horizontal distance of the kite from the base of the string.

Since the height of the kite is fixed at 40 m, then $\tan\theta = \frac{40}{x}$.

Differentiating both sides of this equation with respect to the time t yields

$$\frac{d(\tan\theta)}{dt} = \frac{d}{dt}\left(\frac{40}{x}\right)$$

Applying the chain rule, we obtain

$$\frac{d(\tan\theta)}{d\theta}\frac{d\theta}{dt} = \frac{d\left(\frac{40}{x}\right)}{dx}\frac{dx}{dt} \qquad \text{That is, } \sec^2\theta\frac{d\theta}{dt} = -\frac{40}{x^2}\frac{dx}{dt}$$

We are told that $\frac{dx}{dt} = 2$ m/s, so $\frac{d\theta}{dt} = -\frac{40}{x^2\sec^2\theta}(2) = -\frac{80}{x^2\sec^2\theta}$

When the length of the string is 60 m, then $x = \sqrt{60^2 - 40^2} = 20\sqrt{5}$ so, $\sec\theta = \frac{60}{20\sqrt{5}} = \frac{3}{\sqrt{5}}$.

Therefore, $\frac{d\theta}{dt} = -\frac{80}{2000\left(\frac{9}{5}\right)} = -\frac{1}{45}$ radians/sec or $-\frac{4}{\pi}$ degrees/sec .

That is, when the string is extended to 60 m, the angle of inclination of the kite is decreasing at approximately $4/\pi$ degrees per second.

WORKED EXAMPLE 3

The upper portion of an hour glass is the shape of an inverted cone with radius 5 cm and height 12 cm. When the sand is at a height of 7.2 cm in the upper cone it is draining at a rate of 6 cm^3 /min. How fast is the height of the sand falling from the upper cone at that time?

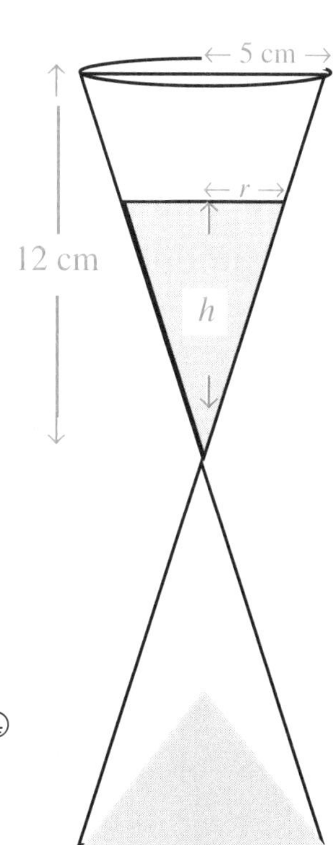

SOLUTION

Proceeding as in *Worked Example* 1, we let V denote the volume of sand in the upper cone, we let r denote the radius (in centimeters) of the cone at a height of h cm above the vertex. Since the angle at the vertex is common to both triangles in the upper cone,

$$\frac{r}{h} = \frac{5}{12} \text{ and so, } r = \frac{5}{12}h \qquad ①$$

The volume of sand in the upper cone is given by $V = \frac{1}{3}\pi r^2 h$. ②

Substituting from equation ① into ② yields $V = \frac{1}{3}\pi\left(\frac{5}{12}\right)^2 h^3$. ③

To relate dV/dt and dh/dt, we differentiate both sides of ③ to obtain $\frac{dV}{dt} = \pi\left(\frac{5}{12}\right)^2 h^2\frac{dh}{dt}$ ④

Substituting $dV/dt = 6$ and $h = 7.2$ into equation 4 and solving for dh/dt yields $dh/dt = 2/(3\pi)$. That is, the height of the sand is falling at about 0.21 cm/min when the height is 7.2 cm.

❶. a) What is the chain rule?
b) Why is the chain rule important?

❷. Use the chain rule to find the derivative of each function with respect to x.

a) $\ln(3x^2+4x)^3$ b) $\dfrac{1}{\sin(5x-1)^2}$

❸. Helium is leaking from a spherical balloon at the rate of 50 cm^3/min. At what rate is the diameter of the balloon shrinking when the balloon has a radius of 6 cm?

❹. The base of a 7-meter long ladder leaning against a vertical wall slides horizontally outward at 50 cm/s. How fast is the top falling when the base is 4 *m* from the wall?

❺. A 6-*m* long ladder is leaned against a wall so that it reaches 5 *m* up the wall. Suddenly it begins to slide down the wall. When the top of the ladder is 3 *m* above the ground, it is moving downward at 10 *cm/s*. At what rate is the foot of the ladder moving away from the wall?

❻. A jet flies west at 150 *m/s* at an altitude of 1000 *m*. A searchlight on the ground in the same vertical plane as the aircraft remains focussed on the plane. What is the rate of revolution (in radians per second) of the searchlight when the jet is 500 *m* due east of the searchlight?

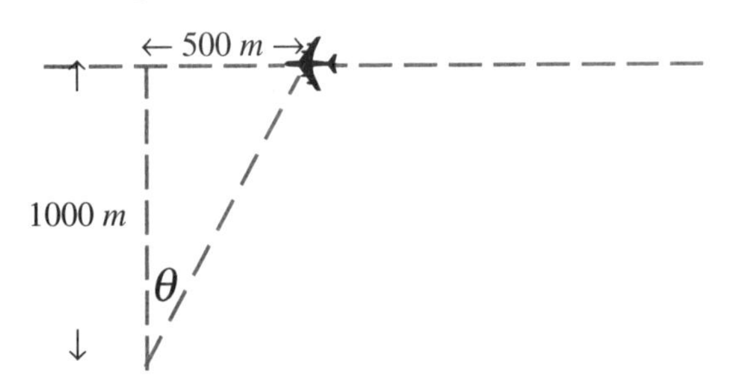

❼. Ship *H. M. S. Coxeter* is sailing due east at 20 km/h. Another ship, the *S. S. Geodesic*, is located 40 km north and 50 km west of the *H. M. S. Coxeter* and is sailing due south at 30 km/h.
a) At what rate are the ships approaching (or departing) one hour later?
b) What is the shortest distance between the two ships?

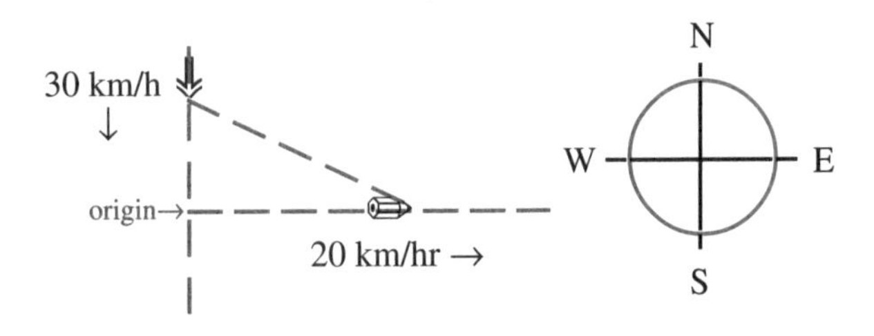

❽. Find the volume of the right circular cylinder of maximum volume which can be inscribed in a sphere of radius R.

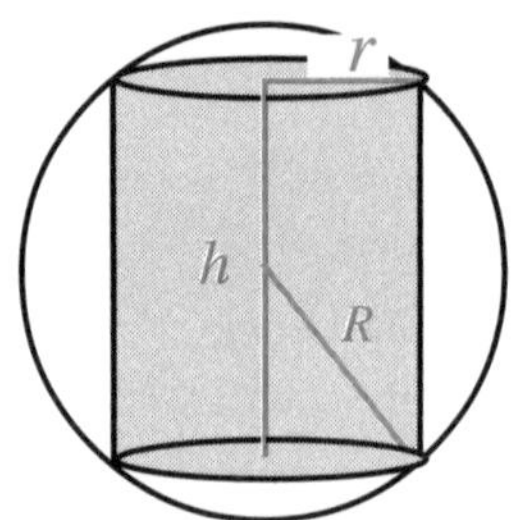

❾. Water is flowing into a cylindrical tank of radius 80 cm at a rate of 0.5 m^3/min. Write an expression which gives the height of the water in the tank as a function of time.

❿. a) Prove from first principles that $\dfrac{dy}{dx} = \dfrac{1}{\left(\frac{dx}{dy}\right)}$
b) Prove part a) using the chain rule.

⓫. The Cartesian equation $b^2x^2 + a^2y^2 = 1$ ($a > b$)defines an ellipse with semi-major axis a and semi-minor axis b. We can also define this ellipse by the *parametric equations*

$$x = a\cos t$$
$$y = b\sin t$$

which express x and y in terms of a common parameter t.
a) Write expressions for dx/dt and dy/dt.
b) Use the chain rule to write an expression for dy/dx.
c) Calculate d^2y/dx^2.
d) Press MODE ▶ ▼ ENTER ENTER to select PARAMETRIC mode. Then graph the ellipse with $a = 5$ and $b = 4$.
e) Calculate dx/dt, dy/dt and dy/dx at $t = 1.2$.
f) Use F5 6 to check your answers in part e).

⓬. A missile with initial velocity V *m/s* (meters per second) and launched at an angle of θ radians follows a trajectory defined by

$$x = (V\cos\theta)\,t$$
$$y = (V\sin\theta)\,t - 4.9t^2$$

where x and y are respectively the horizontal and vertical coordinates of the missile at time t seconds after launch.
a) Calculate dx/dt and dy/dt.
b) Use your answers in part a) to determine dy/dx.
c) At what time will the missile reach the top of its trajectory? (Express your answer in terms of V and θ.)
d) What is the horizontal range of the missile?
e) For what angle θ is the range a maximum?
f) The velocity of the missile at any time t is

$$\frac{ds}{dt} = \sqrt{\left(\frac{dx}{dt}\right)^2 + \left(\frac{dy}{dt}\right)^2}$$

Express the velocity as a function of time.

Unit 4 The Integral — Concepts & Applications

Mathematical Concepts

- •Fundamental Theorem of Calculus
- •the Power Rule of Integration
- •Definite & Indefinite Integrals
- •Formulas for Integrating
  - polynomials
  - algebraic functions
  - exponential and logarithmic functions
  - trigonometric functions
- •Techniques of Integration
  - integration by rational substitution
  - integration by parts
  - integration by trigonometric substitution
  - integration by partial fractions
- •Arc length
- •Areas between curves
- •Surface Areas & Volumes of 3-D Solids

TI-89 Commands

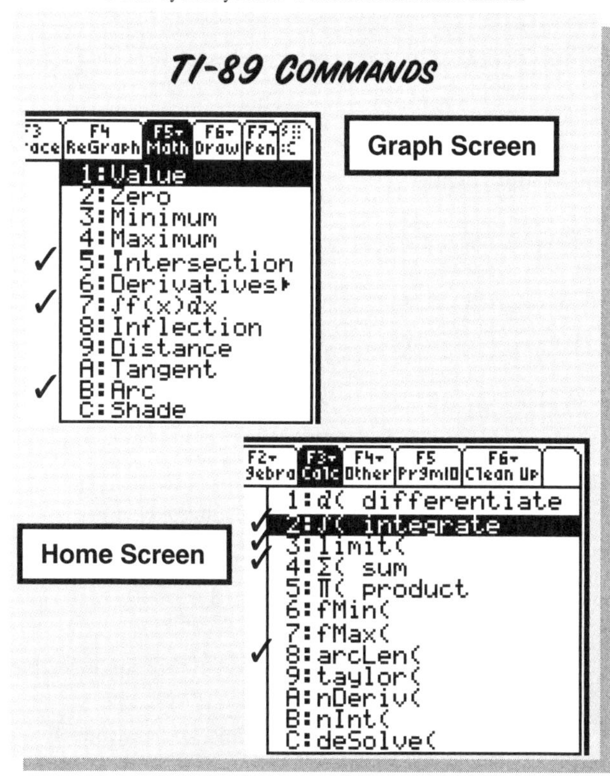

EXPLORATION 14 THE FUNDAMENTAL THEOREM OF CALCULUS

One of Isaac Newton's remarkable achievements was his computation of π to seven decimal places. To achieve this he computed the shaded area of the semi-circle with center (1/2, 0) and radius 1/2 in two different ways.

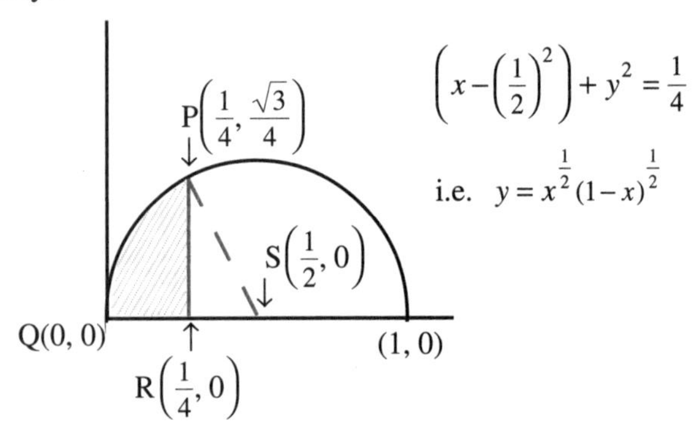

$$\left(x-\left(\frac{1}{2}\right)^2\right)+y^2=\frac{1}{4}$$

i.e. $y = x^{\frac{1}{2}}(1-x)^{\frac{1}{2}}$

OF

ANALYSIS

BY

Equations of an infinite Number of Terms.

1. THE General Method, which I had devifed fome confiderable Time ago, for meafuring the Quantity of Curves, by Means of Series, infinite in the Number of Terms, is rather fhortly explained, than accurately demonftrated in what follows.

2. Let the Bafe AB of any Curve AD have BD for it's perpendicular Ordinate; and call AB=x, BD=y, and let a, b, c, &c. be given Quantities, and m and n whole Numbers. Then

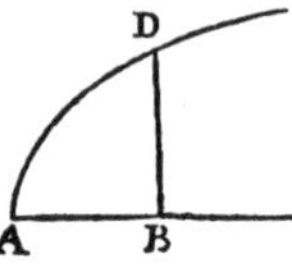

The Quadrature of Simple Curves,

RULE I.

3. If $ax^{\frac{m}{n}}=y$; it fhall be $\frac{an}{m+n}x^{\frac{m+n}{n}}$ =Area ABD.

GEOMETRIC METHOD

area of segment PQR = area of sector SPQ – area of ΔSPR

$$\frac{1}{3}\left(\frac{\pi}{8}\right)-\frac{1}{2}\left(\frac{1}{4}\right)\left(\frac{\sqrt{3}}{4}\right) = \left(\frac{\pi}{24}\right)-\left(\frac{\sqrt{3}}{32}\right)$$

↑ ↑ Area of the semi-circle

∠PSR = π/3

CALCULUS METHOD

To find a formula for the area under the curve $y = x^{\frac{1}{2}}(1-x)^{\frac{1}{2}}$ between 0 and 1/4, Newton used the formula shown in his publication above. That is, the area under the curve defined by $y = ax^{\frac{m}{n}}$ between 0 and x is given by $\frac{an}{m+n}x^{\frac{m+n}{n}}$. In modern notation, we denote the area under the curve defined by $y = f(x)$ and between $x = c$ and $x = d$ (where $d > c$) by the symbol $\int_c^d f(x)dx$ called the *definite integral* of $f(x)$ between c and d. (The integral sign $\int$ is an elongated S which suggests the rectangular sums that we discussed in *Exploration* 3.) In this notation, the area under the curve $y = x^{\frac{1}{2}}(1-x)^{\frac{1}{2}}$ between 0 and 1/4 is denoted by $\int_0^{\frac{1}{4}} x^{\frac{1}{2}}(1-x)^{\frac{1}{2}}dx$.

To evaluate this integral, Newton expanded $(1-x)^{\frac{1}{2}}$ using the binomial theorem which he had developed (see p. 30) and applied the fact that areas are additive. That is,

$$\int_0^{\frac{1}{4}} x^{\frac{1}{2}}(1-x)^{\frac{1}{2}}dx = \int_0^{\frac{1}{4}} x^{\frac{1}{2}}\left(1 - \frac{1}{2}x - \frac{1}{8}x^2 - \frac{1}{16}x^3 - \frac{5}{128}x^4 - \frac{7}{256}x^5 - \ldots\right)dx$$

$$= \int_0^{\frac{1}{4}}\left(x^{\frac{1}{2}} - \frac{1}{2}x^{\frac{3}{2}} - \frac{1}{8}x^{\frac{5}{2}} - \frac{1}{16}x^{\frac{7}{2}} - \frac{5}{128}x^{\frac{9}{2}} - \frac{7}{256}x^{\frac{11}{2}} - \ldots\right)dx$$

$$= \int_0^{\frac{1}{4}} x^{\frac{1}{2}}dx - \int_0^{\frac{1}{4}}\frac{1}{2}x^{\frac{3}{2}}dx - \int_0^{\frac{1}{4}}\frac{1}{8}x^{\frac{5}{2}}dx - \int_0^{\frac{1}{4}}\frac{1}{16}x^{\frac{7}{2}}dx - \int_0^{\frac{1}{4}}\frac{5}{128}x^{\frac{9}{2}}dx - \int_0^{\frac{1}{4}}\frac{7}{256}x^{\frac{11}{2}}dx \ldots \quad ①$$

To evaluate the terms on the right side of this equation, Newton used his formula noted above; i.e., $\int ax^{\frac{m}{n}}dx = \frac{an}{m+n}x^{\frac{m+n}{n}}$, or in modern notation, $\int_0^{\frac{1}{4}} ax^{\frac{m}{n}}dx = \frac{an}{m+n}\left(\frac{1}{4}\right)^{\frac{m+n}{n}}$ ② .

Applying formula ② term-by-term to the right side of equation ① yields

$$\int_0^{\frac{1}{4}} x^{\frac{1}{2}}(1-x)^{\frac{1}{2}}\,dx = \frac{2}{3}\left(\frac{1}{4}\right)^{\frac{3}{2}} - \frac{1}{2}\left(\frac{2}{5}\left(\frac{1}{4}\right)^{\frac{5}{2}}\right) - \frac{1}{8}\left(\frac{2}{7}\left(\frac{1}{4}\right)^{\frac{7}{2}}\right) - \frac{1}{16}\left(\frac{2}{9}\left(\frac{1}{4}\right)^{\frac{9}{2}}\right) - \frac{5}{128}\left(\frac{2}{11}\left(\frac{1}{4}\right)^{\frac{11}{2}}\right) - \frac{7}{256}\left(\frac{2}{13}\left(\frac{1}{4}\right)^{\frac{13}{2}}\right) - \cdots$$

$$= \frac{1}{12} - \frac{1}{160} - \frac{1}{3584} - \frac{1}{36864} - \frac{5}{1441792} - \frac{7}{13631488} - \cdots$$

$$\approx 0.07677310678 \text{ (from the first nine terms of this series)}$$

Equating the areas computed by both methods, Newton wrote, $\left(\frac{\pi}{24}\right) - \left(\frac{\sqrt{3}}{32}\right) \approx 0.07677310678$.
Solving for π yielded $\pi \approx 3.141592668$ which gives the correct value of π to seven decimal places.

The achievement described above represents an intellectual *tour de force.* How did Newton arrive at his formula for the integral of the general power function $y = ax^{\frac{m}{n}}$? In his study of motion, Newton had used fluxions (i.e., derivatives of distance-time functions) to determine instantaneous velocity. Furthermore he had realized that the area under a velocity-time graph between 0 and t represents the total distance traveled. Hence, if $v(x)$ is the velocity of a moving body at time x, and $s(t)$ is the total distance traveled between time 0 and time t, then

$$s(t) = \int_0^t v(x)dx \quad \text{and} \quad \frac{ds(t)}{dt} = v(t), \text{ so} \quad \frac{d\int_0^t v(x)dx}{dt} = v(t)$$

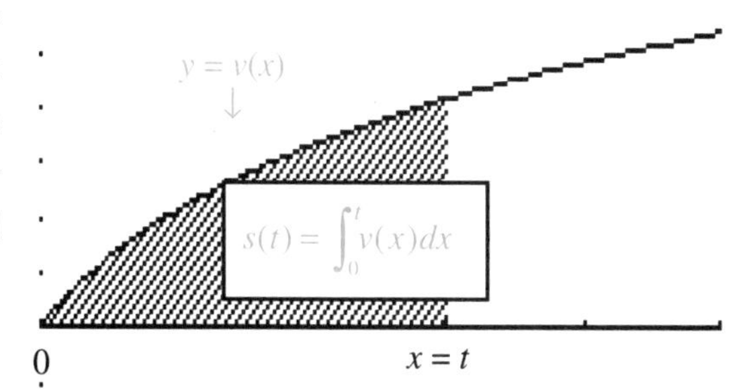

That is, the operations of integration and differentiation are opposite. Newton's realization that an integral is the *anti-derivative* of a function enabled him to obtain a formula for the integral of a power function by reversing the power rule for derivatives. This relationship between the integral of a function and its derivative is called the *Fundamental Theorem of Calculus* because it links differential and integral calculus.

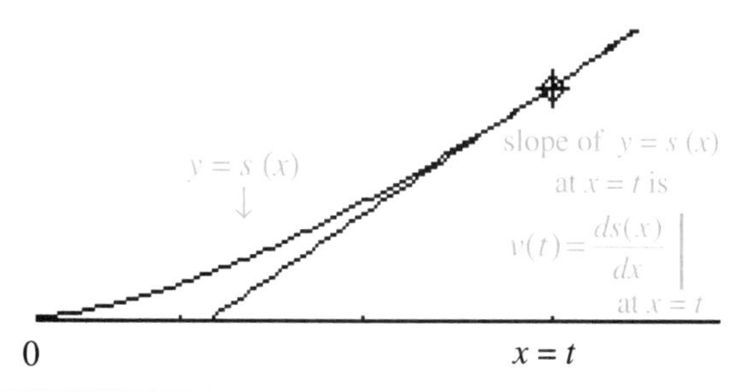

The Fundamental Theorem of Calculus

> If $f(x)$ is a continuous function on an interval $[a, b]$, then $\int_a^x f(t)dt$ is a differentiable function on $[a, b]$ and $\frac{d\int_a^x f(t)dt}{dx} = f(x)$.

Proof

Let $F(x) = \int_a^x f(t)dt$. Then $\frac{dF(x)}{dx} = \lim_{\Delta x \to 0} \frac{F(x+\Delta x) - F(x)}{\Delta x} = \lim_{\Delta x \to 0} \frac{\int_a^{x+\Delta x} f(t)dt - \int_a^x f(t)dt}{\Delta x}$

The area under the curve $y = f(t)$ between $t = a$ and $t = x + \Delta x$ less the area between $t = a$ and $t = x$, is the area between x and $x + \Delta x$.

$$= \lim_{\Delta x \to 0} \frac{\int_x^{x+\Delta x} f(t)dt}{\Delta x} = \lim_{\Delta x \to 0} \frac{f(x+\theta\Delta x)\Delta x}{\Delta x} \quad \text{where } 0 \le \theta \le 1$$

$$= f(x)$$

The area under the curve $y = f(t)$ between x and $x + \Delta x$ is the "average" value of $f(x)$ over that interval times Δx.

The function $F(x)$ defined by $F(x) = \int_a^x f(t)dt$ is said to be an antiderivative of $f(x)$ because differentiating this function with respect to x yields $f(x)$.

WORKED EXAMPLES

Suppose $F_1(x)$ and $F_2(x)$ are two different antiderivatives of $f(x)$; i.e.,

$$\frac{dF_1(x)}{dx} = \frac{dF_2(x)}{dx} = f(x) \quad \text{then} \quad \frac{d\big(F_1(x) - F_2(x)\big)}{dx} = f(x) - f(x) = 0.$$

Therefore, all antiderivatives of $f(x)$ differ by a constant.

Definition

> If $f(x)$ is a continuous function, $\int f(x)dx$ denotes the general antiderivative of $f(x)$ from which all antiderivatives of $f(x)$ can be obtained by the addition of a constant. $\int f(x)dx$ is called the *indefinite integral* of $f(x)$.

From the Fundamental Theorem of Calculus, we can deduce the following corollary.

Corollary

> If $f(x)$ is a continuous function and $F'(x) = f(x)$, then $\int_a^b f(x)dx = F(b) - F(a)$

We can use the Fundamental Theorem of Calculus to prove the following *Power Rule of Integration,* which is the rule Newton used in his computation of π. (See investigation ⑩ p. 65)

The Power Rule of Integration

> $\int x^n dx = \frac{x^{n+1}}{n+1} + C$ for n rational but $n \neq 1$.

WORKED EXAMPLE

Evaluate the following integrals. a) $\int_0^4 x^5 dx$ b) $\int_0^1 \sqrt{x}\, dx$ c) $\int_1^4 \frac{1}{\sqrt[3]{x+4}} dx$

SOLUTION

a) Application of the power rule yields $F(x) = \int x^5 dx = \frac{x^6}{6} + C$. Using the Corollary above, we obtain

$$\int_0^4 x^5 dx = F(4) - F(0)$$
$$= \left(\frac{4^6}{6} + C\right) - \left(\frac{0^6}{6} + C\right) = \frac{4^6}{6} \text{ or } \frac{2048}{3}$$

This expression is written $\left.\frac{x^6}{6}\right|_0^4$

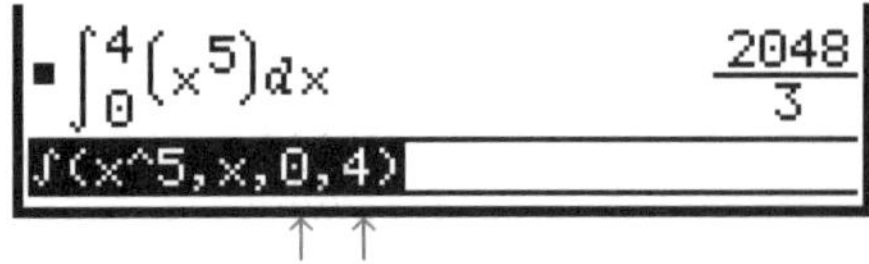

What do we get if we leave out the limits of integration?

$\int_0^1 \sqrt{x}\, dx$ 2/3

∫(√(x),x,0,1)

To check this answer on the TI-89, press **F3** **2** and then enter the integrand, the variable and the limits of integration as shown in the command line of the top display.

b) Application of the Power Rule and the corollary of the Fundamental Theorem of Calculus yields.

$$\int_0^1 \sqrt{x}\, dx = \int_0^1 x^{\frac{1}{2}} dx = \left.\frac{2}{3}x^{\frac{3}{2}}\right|_0^1 = \frac{2}{3}$$

The middle display shows how to verify our answer on the TI-89.

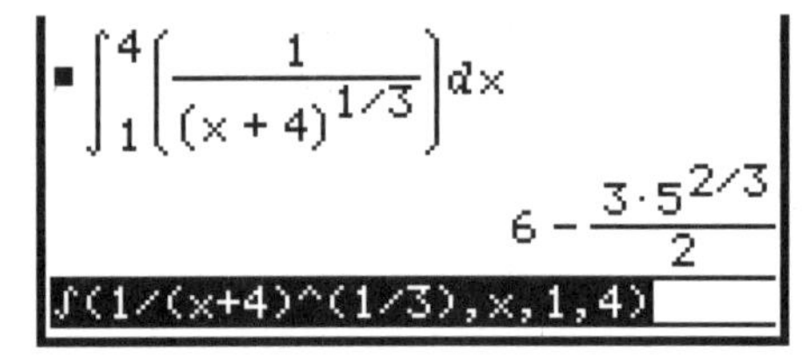

c) Application of the Power Rule and the corollary of the Fundamental Theorem of Calculus yields.

$$\int_1^4 \frac{1}{\sqrt[3]{x+4}} dx = \int_1^4 (x+4)^{-\frac{1}{3}} dx = \left.\frac{3}{2}(x+4)^{\frac{2}{3}}\right|_1^4 = \frac{3}{2}\left[8^{\frac{2}{3}} - 5^{\frac{2}{3}}\right] = 6 - \frac{3}{2}5^{\frac{2}{3}}$$

The bottom display shows how we can verify our answer with the TI-89.

Exercises & Investigations

1. a) Explain the meaning of the Fundamental Theorem of Calculus.

b) Explain how we can use this theorem to integrate

$$\int_0^x \sqrt{t}\,dt$$

2. a) Evaluate $\int_{-1}^{1}(x^3 - x)dx$.

b) Graph the function $y = x^3 - x$ in the window

$-1.5 \le x \le 1.5;\ -1 \le y \le 1$

Press **F5** **7** when displaying the graph. Then enter – 1 and 1 respectively in response to the prompts **Lower Limit?** and **Upper Limit?** Observe the value of the integral in the bottom left corner of the display. Does this mean that the total area bounded by the graph and the x-axis is zero?

c) Use the procedure in part b) to evaluate

$$\int_{-1}^{0}(x^3 - x)dx \text{ and } \int_{0}^{1}(x^3 - x)dx$$

d) Explain what is meant by the term *signed area.*

3. Interpreting $\int_a^b f(x)dx$ as the (signed) area between $x = a$ and $x = b$ and under the curve defined by $y = f(x)$, explain why each of the following identities is true.

a) $\int_a^b f(x)dx = \int_0^b f(x)dx - \int_0^a f(x)dx$

b) $\int_a^b f(x)dx = -\int_b^a f(x)dx$

4. a) Find each indefinite integral.

(i) $\int 3x^2dx$ (ii) $\int (x+4)^{\frac{3}{2}}dx$ (iii) $\int \frac{1}{(x+5)^2}dx$

b) Use your results in part a) to evaluate each integral.

(i) $\int_0^4 3x^2dx$ (ii) $\int_{-1}^{3} (x+4)^{\frac{3}{2}}dx$ (iii) $\int_1^5 \frac{1}{(x+5)^2}dx$

c) Use your TI-89 to verify your answers in parts a) and b).

d) Describe the region for which each integral in part b) gives the area.

5. Integrate $\int \left(\frac{x^3+x+1}{x^3}\right)dx$. Evaluate $\int_{-1}^{1}\left(\frac{x^3+x+1}{x^3}\right)dx$

Note: The integral of a sum is the sum of the integrals. Why?

6. Integrate $\int (1-x)\sqrt{x}\,dx$. Evaluate $\int_0^6 (1-x)\sqrt{x}\,dx$

7. a) Evaluate using the method of exhaustion (p. 18).

(i) $\int_0^5 x^2dx$ (ii) $\int_0^1 (1-x^2)dx$ (iii) $\int_0^2 x(1-x)dx$

b) Use your TI-89 to check your answers in part a).

8. a) Prove that if $f(x)$ is a differentiable function of x, then

$$\int [f(x)]^n\left(\frac{d\,f(x)}{dx}\right)dx = \frac{[f(x)]^{n+1}}{n+1} + C$$

Hint: Let $f(x) = u$ and apply the power rule for integrals to u.

b) Use the generalized power rule in part a) to compute these indefinite integrals.

(i) $\int \frac{2x+1}{(x^2+x+3)^2}dx$ (ii) $\int \sqrt{x^2+5x}(2x+5)dx$

(iii) $\int \frac{x^2}{\sqrt{1-x^3}}dx$ (iv) $\int \frac{2x+1}{\sqrt{x^2+x+3}}dx$

c) Use your TI-89 to check your answers in part b).

9. On page 18, we used the method of exhaustion to find the area of the first quadrant bounded by the parabola with equation $y = -x^2 + 4$.

a) Graph this equation and evaluate the area by selecting the $\int f(x)dx$ command on the F5 menu.

b) Evaluate this area using **F3** **2** and entering the function on the command line of the home screen.

10. Prove the Power Rule of Integration

$$\int x^n dx = \frac{x^{n+1}}{n+1} + C \quad \text{for } n \text{ rational but } n \ne 1.$$

by differentiating each side with respect to x.

11. Use the Fundamental Theorem of Calculus and the rules for differentiating exponential, logarithmic and trig functions to derive formulas for each integral.

a) $\int e^x dx$ b) $\int \frac{1}{x}dx$ c) $\int \sin x\,dx$ d) $\int \cos x\,dx$

12. Derive a formula for the horizontal distance s, traveled by a projectile in t seconds after launch if it accelerates at a constant rate of -3 *m/s*^{2}, and its initial velocity is 60 *m/s.* Use your formula to calculate the distance traveled when its velocity reaches 0.

13. a) Use the method of exhaustion (see p. 18) to express

$\int_0^1 \sqrt{t}\,dt$ as the limit of a sum.

b) Use the power rule for integration to evaluate the integral in part a).

c) Use your answer in part b) to show that

$$\lim_{n\to\infty}\frac{1}{n}\sum_{k=1}^{n}\sqrt{\frac{k}{n}} = \frac{2}{3}$$

d) Follow the same procedures as in parts a), b) and c) to show that

$$\lim_{n\to\infty}\frac{1}{n}\sum_{k=1}^{n}\sqrt{1-\frac{k}{n}} = \frac{2}{3}$$

EXPLORATION 15 INTEGRALS OF EXPONENTIAL, LOG, AND TRIG FUNCTIONS

In *Worked Example* 4 on page 23, we modeled the world population since 1900 A.D. with the logistic function $P(t)$ where $P(t)$ is given by

$$P(t) = \frac{78.12}{6.3 + 102e^{-0.02817t}}$$

Where did this formula come from? You will discover the answer in *Worked Example* 3 p. 68, but first we must develop some important integration formulas which you will need.

In *Exploration* 10, we discovered the formula

$$\frac{d\, e^{f(x)}}{dx} = e^{f(x)}\left(\frac{d\, f(x)}{dx}\right)$$

This means that if $F(x) = e^{f(x)}$, then $F'(x) = e^{f(x)}\left(\frac{d\, f(x)}{dx}\right)$.

Hence it follows from the Fundamental Theorem of Calculus that $F(x)$ is an indefinite integral of $e^{f(x)}\left(\frac{d\, f(x)}{dx}\right)$. That is,

$$\int e^{f(x)}\left(\frac{d\, f(x)}{dx}\right)du = e^{f(x)} + C$$

We write this in the following more compact form

$$\boxed{\int e^u\, du = e^u + C}$$

← where u is any function for which the integral is defined

The replacement of $f(x)$ by u is conventional notation introduced to simplify the formulas and make their structure more apparent. Proceeding as above, we can apply our formulas for derivatives in reverse to obtain the following formulas.

$$\boxed{\int a^u\, du = \frac{a^u}{\ln a} + C}$$

← where a is any constant

$$\boxed{\int \frac{1}{u}\, du = \ln|u| + C}$$

$$\boxed{\int \sin u\, du = -\cos u + C}$$

$$\boxed{\int \cos u\, du = -\sin u + C}$$

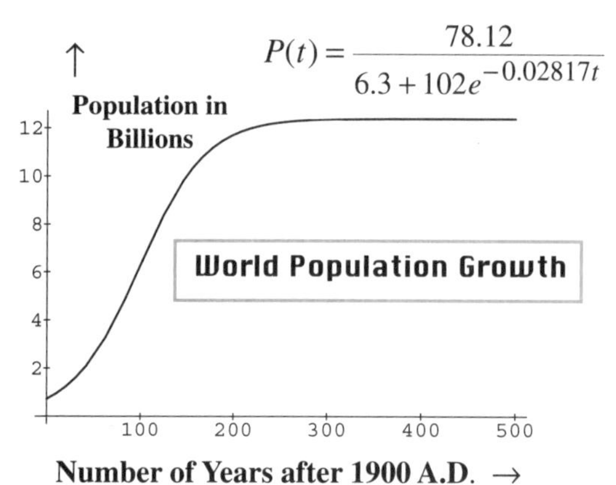

We can combine the formulas for $\int \frac{1}{u} du$ and $\int \cos u\, du$ to obtain $\int \tan u\, du$ as follows

$$\int \tan u\, du = \int \frac{\sin u}{\cos u}\, du = -\int \frac{d(\cos u)}{\cos u} = -\ln|\cos u| + C$$

That is,

$$\boxed{\int \tan u\, du = -\ln|\cos u| + C}$$

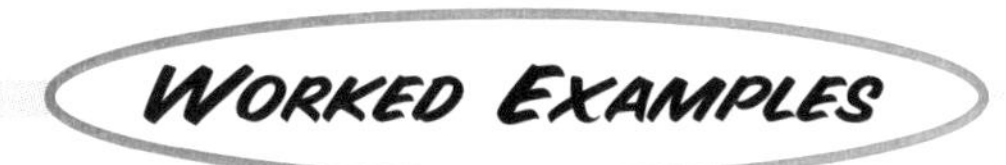

WORKED EXAMPLE 1

Integrate a) $\int e^{3x+5}\,dx$ b) $\int \frac{dx}{4x+5}$ c) $\int \left(\frac{3}{4}\right)^x dx$ d) $\int -\sin\frac{x}{2}dx$

Use your TI-89 to verify your answers.

SOLUTION

a) We set $u = 3x + 5$, so $du = 3dx$. Substitution yields

$$\int e^{3x+5}\,dx = \int e^u \frac{du}{3} = \frac{1}{3}e^u + C = \frac{1}{3}e^{3x+5} + C$$

To verify this answer on your TI-89, enter the commands as shown on the command line of the display on the right . The output agrees with the answer presented above.

$$\blacksquare \int \left(e^{3\cdot x+5}\right)dx \qquad \frac{e^{3\cdot x+5}}{3}$$

```
∫(e^(3x+5),x)
```

b) We set $u = 4x + 5$, so $du = 4dx$. Substitution yields

$$\int \frac{dx}{4x+5} = \int \frac{du}{u} = \frac{1}{4}\ln u + C = \frac{1}{4}\ln|4x+5| + C$$

To verify this answer on your TI-89, enter the commands as shown on the command line of the display on the right .

$$\blacksquare \int \left(\frac{1}{4\cdot x+5}\right)dx \qquad \frac{\ln(|4\cdot x+5|)}{4}$$

```
∫(1/(4x+5),x)
```

c) To evaluate $\int \left(\frac{3}{4}\right)^x dx$ we can apply the exponential formula with $a = 3/4$. This yields

$$\int \left(\frac{3}{4}\right)^x dx = \frac{\left(\frac{3}{4}\right)^x}{\ln(3/4)}$$

$$\blacksquare \int \left((3/4)^x\right)dx \qquad \frac{(3/4)^x}{\ln(3/4)}$$

```
∫((3/4)^x,x)
```

d) To evaluate $\int -\sin\frac{x}{2}dx$ we set $u = x/2$, then $du = dx/2$ so $dx = 2du$.

$$-\int \sin\frac{x}{2}\,dx = -\int \sin u\,(2du) = -2\int \sin u\,du = 2\cos\frac{x}{2} + C$$

> Sometimes your TI-89 will display an integral in a different form from the one you compute. If you cannot show the their equivalence algebraically, graph the computed answer plus 1 minus the TI-89 answer. The resulting graph should be the line $y = 1$.

$$\blacksquare \int -\sin\left(\frac{x}{2}\right)dx \qquad 2\cdot\cos\left(\frac{x}{2}\right)$$

```
∫(-sin(x/2),x)
```

WORKED EXAMPLES

WORKED EXAMPLE 2

Integrate a) $\int \sin x \cos x \, dx$ b) $\int \frac{6x+2}{3x^2+2x+1} dx$ c) $\int \sin x \, e^{\cos x} \, dx$.

Use your TI-89 to verify your answers.

SOLUTION

a) $\int \sin x \cos x \, dx = \frac{1}{2}\int 2\sin x \cos x \, dx$

$= \frac{1}{2}\int \sin 2x \, dx = \frac{1}{4}\int \sin 2x \, d(2x)$

$= -\frac{1}{4}\cos 2x + C$ Alternatively, $\int \sin x \cos x \, dx = \int u \, du = \frac{u^2}{2} + C = \frac{\sin^2 x}{2} + C$ (where $u = \sin x$)

```
■ ∫(sin(x)·cos(x))dx
                    -(cos(x))^2 / 2
∫(sin(x)cos(x),x)
```

b) $\int \frac{6x+2}{3x^2+2x+1} dx = \int \frac{du}{u} = \ln|u| + C = \ln\left|3x^2+2x+1\right| + C$ (where $u = 6x+2$)

```
■ ∫((6·x + 2)/(3·x^2 + 2·x + 1))dx
                    ln(|3·x^2 + 2·x + 1|)
∫((6x+2)/(3x^2+2x+1),x)
```

c) $\int \sin x \, e^{\cos x} \, dx = - \int e^u \, du = -e^u + C = -e^{\cos x} + C$ (where $u = \cos x$)

```
■ ∫(sin(x)·e^cos(x))dx
                    -e^cos(x)
∫(sin(x)e^(cos(x)),x)
```

WORKED EXAMPLE 3

Suppose the fertility rate (i.e., the number of offspring per individual) of a population $P(t)$ is given by the linear function $a - bP(t)$ where a and b are constants. Show that $P(t)$ is the logistic function defined by

$$P(t) = \frac{a}{b + ce^{-at}}$$ where c is a positive constant.

SOLUTION

The change in the population $\frac{dP(t)}{dt}$ is the product of the fertility rate and the population.

That is, $\frac{dP(t)}{dt} = (a - bP(t))\,P(t)$.

Collecting $P(t)$ yields $\frac{dP(t)}{(a-bP(t))\,P(t)} = dt$, and so $\int \frac{dP(t)}{(a-bP(t))\,P(t)} = \int dt$. ①

We express the left side of ① as a sum of partial fractions and integrate as follows.

$$\int\left(\frac{1}{aP(t)} + \frac{b}{a(a-bP(t))}\right)dP(t) = \int dt \quad \Rightarrow$$

$$\int \frac{1}{aP(t)}dP(t) + \int \frac{b}{a(a-bP(t))}dP(t) = \int dt$$

The display shows how we can use the **expand(** command to express the left side of ① as a sum of partial fractions.

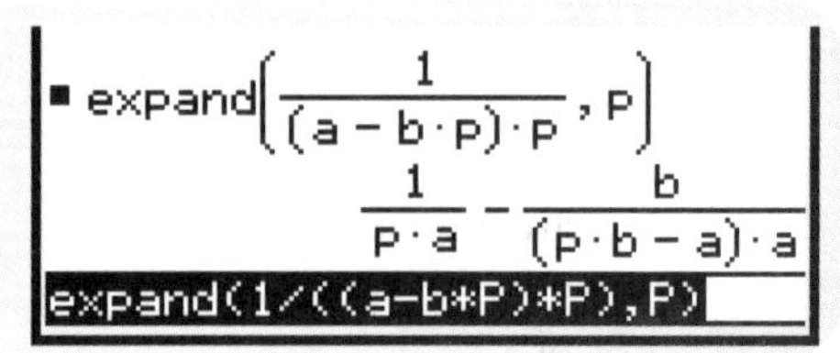

Integrating both sides yields $\frac{1}{a}\ln P(t) - \frac{1}{a}\ln(a - bP(t)) = t + C$. ②

Simplifying equation ② we obtain $\ln\left(\frac{P(t)}{a-bP(t)}\right) = a(t+C)$ and so, $P(t) = \frac{a}{b + ce^{-at}}$ where $c = e^{-aC}$

Exercises & Investigations

1. Explain how the Fundamental Theorem of Calculus is used to obtain formulas for the indefinite integrals of various functions.

2. $F(x)$ and $G(x)$ are two differentiable functions such that

$$\frac{d\,F(x)}{dx} = f(x) \text{ and } \frac{d\,G(x)}{dx} = f(x)$$

Is $F(x) = G(x)$? Explain.

3. In *Worked Example* 2a), we found two different indefinite integrals and the TI-89 display shows a third. Prove that all three integrals differ at most by a constant.

4. Integrate.

a) $\int e^{-3x}\,dx$ b) $\int 2^x dx$ c) $\int xe^{x^2}dx$

5. Use your answers in exercise **4** to evaluate each integral.

a) $\int_0^1 e^{-3x}dx$ b) $\int_0^5 2^x dx$ c) $\int_1^2 xe^{x^2}dx$

6. Integrate.

a) $\int \frac{2}{x}dx$ b) $\int \frac{2x}{x^2+1}dx$ c) $\int \frac{x+1}{2x^2+4x+3}dx$

7. Use your answers in exercise **6** to evaluate each integral.

a) $\int_1^2 \frac{2}{x}dx$ b) $\int_0^3 \frac{2x}{x^2+1}dx$ c) $\int_{-1}^5 \frac{x+1}{2x^2+4x+3}dx$

Describe the region for which each integral gives the area.

8. Integrate.

a) $\int \frac{2}{x\ln x}dx$ b) $\int \sin^2 x\cos x\,dx$ c) $\int \frac{\sin x}{\cos^3 x}dx$

9. Use $\sin^2 x = \frac{1-\cos 2x}{2}$ and /or $\cos^2 x = \frac{1+\cos 2x}{2}$ to calculate each integral.

a) $\int \sin^2 x\,dx$ b) $\int \cos^2 x\,dx$ c) $\int \sin^3 x\,dx$

10. Integrate.

a) $\int \frac{e^{\frac{1}{x}}}{x^2}dx$ b) $\int \frac{dx}{e^x+1}$ c) $\int \frac{\sin x+\cos x}{\cos x}dx$

11. Graph each equation in the given window. Then calculate the area above the x-axis bounded by the graph between the given values of x.

a) $y = \sin^2 x$ $\quad 0 \le x \le \pi;\ -2 \le y \le 2$
b) $y = e^x$ $\quad -1 \le x \le 4;\ -10 \le y \le 60$
c) $y = \frac{1}{(x+2)}$ $\quad -1 \le x \le 5;\ -1 \le y \le 1$

12. Display each graph in exercise **11**. From the graph screen, press **F5** **7** and enter the appropriate lower and upper bounds for x in response to the prompts. Compare your answers in exercises **11** and **12**.

13. In the final investigation of *Exploration* 2 (p. 15), we explored the harmonic series

$$1+\frac{1}{2}+\frac{1}{3}+\frac{1}{4}+\ldots+\frac{1}{n}+\ldots$$

We sought $\lim_{n\to\infty} S(n)$ where $S(n)=\sum_{k=1}^{n}\frac{1}{k}$

Leibniz died believing $\lim_{n\to\infty}\sum_{k=1}^{n}\frac{1}{k} < \infty$. Was he correct?

Check your notes to see what you conjectured based on your computation of $S(100)$, $S(1000)$, and $S(5000)$.

a) Consider an upper rectangular approximation to the area between the x-axis and the graph of the equation $y = 1/x$ in the interval $1 \le x \le n$.

Total area of the rectangles > area under the curve $y = 1/x$.

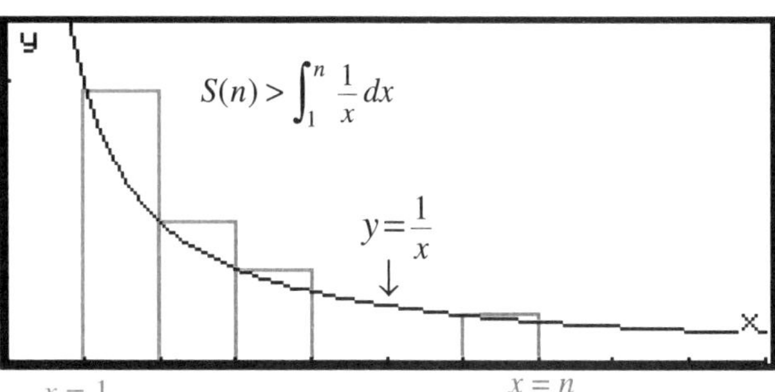

Express $\int_1^n \frac{1}{x}dx$ as a function of n.

b) Explain why $\lim_{n\to\infty} S(n) > \int_1^n \frac{1}{x}dx$.

c) Evaluate the right side of the inequality in part b). Use your answer to determine whether your conjecture was right.

14. In *Exploration* 16, we develop techniques for integrating a wider variety of functions. However, most functions are not integrable. We can only integrate such functions over specific intervals using numerical approximations. A famous non-integrable function is the frequency function for the normal distribution,

$$f(x)=\frac{1}{\sqrt{2\pi}}e^{-\frac{x^2}{2}}$$

a) What happens when you seek the indefinite integral using this command on your TI-89?

```
∫((1/√(2π))e^(-x^2/2),x)
```

b) $\frac{1}{\sqrt{2\pi}}\int_a^b e^{-x^2/2}dx$ gives the probability that a normally distributed variable x takes a value in the interval $a \le x \le b$. Determine the probability associated with each interval

(i) $-\infty \le x \le \infty$ (ii) $0 \le x \le \infty$ (iii) $-1 \le x \le 1$

EXPLORATION 16 INTEGRATION TECHNIQUES

In *Exploration* 14, we studied the ingenious method that Newton employed to calculate π to seven decimal places. To calculate the area under the semi-circle defined by the equation

$$y = x^{\frac{1}{2}}(1-x)^{\frac{1}{2}}$$

and between $x = 0$ and $x = 1/4$, it was necessary to calculate $\int_0^{\frac{1}{4}} x^{\frac{1}{2}}(1-x)^{\frac{1}{2}}\,dx$. To do this, he expanded the binomial $(1-x)^{\frac{1}{2}}$ using his binomial theorem for rational exponents. He then evaluated the expansion term-by-term and added the first nine terms to approximate the infinite series. We might ask whether it is possible to calculate the indefinite integral of y and then substitute the limits 0 and 1/4 to evaluate the definite integral above. A technique for doing this will be developed presently.

In the century that followed the invention of calculus, the foundations of the subject were placed on a more rigorous foundation. Techniques of integration were developed and streamlined, and integrands were classified into special cases according to the substitutions which would transform them into integrable form. The integrand $x^{\frac{1}{2}}(1-x)^{\frac{1}{2}}$ is one such integral which yields to the trigonometric substitution $x^{1/2} \to \sin\theta$.

We substitute $x^{\frac{1}{2}} = \sin\theta$, so $\frac{1}{2}x^{-\frac{1}{2}}dx = \cos\theta\, d\theta$ i.e. $dx = 2x^{\frac{1}{2}}\cos\theta\, d\theta = 2\sin\theta\cos\theta\, d\theta$

Substituting for $x^{\frac{1}{2}}$ and dx in the integrand yields

$$\int x^{\frac{1}{2}}(1-x)^{\frac{1}{2}}dx = \int \underset{x^{1/2}}{\sin\theta}\,\underset{(1-x)^{1/2}}{(1-\sin^2\theta)^{\frac{1}{2}}}\,\underset{dx}{(2\sin\theta\cos\theta\, d\theta)} = \int \sin\theta\cos\theta(2\sin\theta\cos\theta\, d\theta)$$ because $1 - \cos^2\theta = \sin^2\theta$

$$= \frac{1}{2}\int \sin^2 2\theta\, d\theta$$ because $\sin 2\theta = 2\sin\theta\cos\theta$

$$= \frac{1}{4}\int \sin^2 2\theta\, d(2\theta)$$ changing $d\theta$ to $d(2\theta)$

$$= \frac{1}{4}\int\left(\frac{1-\cos 4\theta}{2}\right)d(2\theta)$$ because $\cos 4\theta = 1 - 2\sin^2 2\theta$

$$= \frac{1}{16}\int(1-\cos 4\theta)\,d(4\theta)$$ changing $d(2\theta)$ to $d(4\theta)$

$$= \frac{4\theta}{16} - \frac{\sin 4\theta}{16} + C$$ integrating term-by-term with respect to 4θ.

$$= \frac{\theta}{4} - \frac{1}{16}(4\sin\theta\cos\theta)(1-2\sin^2\theta) + C$$ applying the double angle formulas

$$= \frac{\sin^{-1}\sqrt{x}}{4} - \frac{\sqrt{x}\sqrt{1-x}}{4} + \frac{x^{\frac{3}{2}}\sqrt{1-x}}{2} + C$$ substituting for $\sin\theta$

Whew!

Our trigonometric substitution $x^{1/2} \to \sin\theta$ transforms the integral into an expression in sines and cosines which can be integrated. After the integration, we apply the inverse substitution $\sin\theta \to x^{1/2}$ to express our indefinite integral as a function of x. Terms which involve $\sin\theta$ and $\cos\theta$ become algebraic functions of x, but terms involving θ must be expressed as $\sin^{-1} x^{1/2}$ (read as *the angle whose sine is* $x^{1/2}$). To evaluate such terms on your TI-89 for specific values of x, use

2nd [$\sin^{-1}$]

We can now evaluate $\int_0^{\frac{1}{4}} x^{\frac{1}{2}}(1-x)^{\frac{1}{2}}dx$ by substituting the upper and lower limits for x.

$$\int_0^{\frac{1}{4}} x^{\frac{1}{2}}(1-x)^{\frac{1}{2}}dx = \left.\frac{\sin^{-1}\sqrt{x}}{4}\right|_0^{\frac{1}{4}} - \left.\frac{\sqrt{x}\sqrt{1-x}}{4}\right|_0^{\frac{1}{4}} + \left.\frac{x^{\frac{3}{2}}\sqrt{1-x}}{2}\right|_0^{\frac{1}{4}}$$

$$= \frac{\pi}{24} - \frac{\sqrt{3}}{16} + \frac{\sqrt{3}}{32}$$

$$= \frac{\pi}{4} - \frac{\sqrt{3}}{32}$$

We observe that the value of this definite integral is the same as the value found by Newton using elementary geometry. However, the computation of the indefinite integral solves a much more general problem since it yields the area under the semi-circle between *any* two values of x.

EXERCISES & INVESTIGATIONS

In the integration of $\int \sqrt{x}\sqrt{1-x}\,dx$ the substitution $\sqrt{x} = \sin\theta$ was suggested by the presence of the $\sqrt{1-x}$ term which becomes $\cos\theta$ when we apply the identity $\cos^2\theta = 1 - \sin^2\theta$. When a " + " sign appears under the radical, it suggests a substitution of the form $x = \tan\theta$ because $\sec^2\theta = 1 + \tan^2\theta$.

WORKED EXAMPLE 1

Integrate $\int \frac{dx}{x^2\sqrt{9+x^2}}$.

SOLUTION

Let $x = 3\tan\theta$. Then $dx = 3\sec^2\theta\, d\theta$. Note: $\frac{d}{d\theta}\tan\theta = \sec^2\theta$

$$\int \frac{dx}{x^2\sqrt{9+x^2}} = \int \frac{3\sec^2\theta\, d\theta}{9\tan^2\theta\sqrt{9+9\tan^2\theta}}$$

$$= \frac{1}{9}\int \frac{\sec^2\theta\, d\theta}{\tan^2\theta\sqrt{1+\tan^2\theta}}$$

$$= \frac{1}{9}\int \frac{\cos\theta\, d\theta}{\sin^2\theta} \quad \text{because } \sec^2\theta = 1 + \tan^2\theta$$

$$= \frac{-1}{9\sin\theta} + C \quad = -\frac{\sqrt{x^2+9}}{9x} + C$$

A diagram like the one below is often helpful in expressing the trig functions of θ in terms of x.

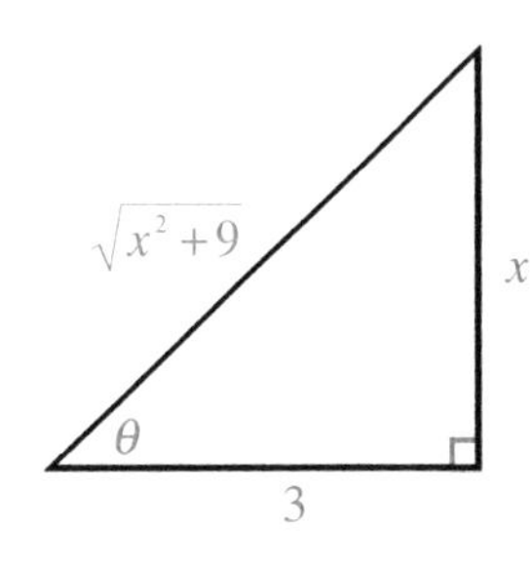

$x = 3\tan\theta$ so, the hypotenuse is $\sqrt{x^2+9}$

$\sin\theta = \frac{x}{\sqrt{x^2+9}}$

One of the most powerful integration techniques is called *integration by parts.* This technique uses an identity which we obtain by integrating the product rule for derivatives. The product rule for the derivative of the product of two differentiable functions $u(x)$ and $v(x)$ is

$$\frac{d}{dx}u(x)v(x) = u(x)\frac{d}{dx}v(x) + v(x)\frac{d}{dx}u(x)$$

Integrating both sides with respect to x yields $\int \frac{d}{dx}u(x)v(x)\,dx = \int u(x)\frac{d}{dx}v(x)\,dx + \int v(x)\frac{d}{dx}u(x)\,dx$

Applying the Fundamental Theorem of Calculus to the left side of the equation yields $u(x)v(x) = \int u(x)\frac{d}{dx}v(x)\,dx + \int v(x)\frac{d}{dx}u(x)\,dx$

Isolating $\int u(x)\frac{d}{dx}v(x)\,dx$ yields the identity

Integration by Parts Identity $\quad \int u(x)\frac{d}{dx}v(x)\,dx = u(x)v(x) - \int v(x)\frac{d}{dx}u(x)\,dx$

To apply integration by parts, we write the integrand as a product of a differentiable function $u(x)$ and a function which is integrable which we represent as $\frac{d}{dx}v(x)$. Then we calculate $v(x)$ and $\frac{d}{dx}u(x)$ and substitute into the right side of the identity. We then attempt to integrate the right side of the identity to obtain an expression for the integral on the left side.

This technique does not always yield a solution, although a clever choice of $u(x)$ and $v(x)$ sometimes yields a solution to an otherwise difficult integral.

Worked Examples

Integration by parts often requires an insightful (or lucky) choice of $u(x)$ and $\frac{d}{dx}v(x)$ such that $v(x)\frac{d}{dx}u(x)$ is integrable. The following example shows a successful application of integration by parts.

Worked Example 2

Integrate $\int x^3 e^{-x^2}\,dx$. Use your TI-89 to verify your answer.

Solution

Choose $u = x^2$, then $\frac{d}{dx}v(x) = xe^{-x^2}$. Also $du = 2x\,dx$ and $v(x) = \int xe^{-x^2}\,dx = -\frac{1}{2}e^{-x^2}$.

Substituting these expressions for $u(x)$ and $v(x)$ into the integration by parts identity yields

$$\int x^3 e^{-x^2}\,dx = u(x)v(x) - \int v(x)\frac{du(x)}{dx}dx$$

$$= -\frac{x^2}{2}e^{-x^2} + \int xe^{-x^2}\,dx$$

$$= -\frac{x^2}{2}e^{-x^2} - \frac{1}{2}e^{-x^2} + C$$

$$= -\frac{1}{2}(1+x^2)e^{-x^2} + C$$

The display shows the command and the output that verify our computations.

$$\blacksquare \int \left(x^3 \cdot e^{-x^2}\right) dx \qquad \frac{-\left(x^2+1\right)\cdot e^{-x^2}}{2}$$

```
∫(x^3e^(-x^2),x)
```

The technique used in the next example is called *integration by partial fractions.* It is often useful when the integrand is a quotient of two polynomials.

Worked Example 3

Integrate $\int \frac{2x^3+4x+10}{x^4+x^2+2x}dx$. Use your TI-89 to verify your answer.

Solution

Using algebra or the TI-89, we can express the quotient in the integrand as a sum of partial fractions.

$$\frac{2x^3+4x+10}{x^4+x^2+2x} = \frac{-2x+1}{x^2-x+2} - \frac{1}{x+1} + \frac{5}{x} \quad \text{and so,}$$

$$\int \frac{2x^3+4x+10}{x^4+x^2+2x}dx = \int \frac{-2x+1}{x^2-x+2}dx - \int \frac{1}{x+1}dx + \int \frac{5}{x}dx$$

$$= -\ln\left|x^2-x+2\right| - \ln|x+1| + 5\ln|x| + C$$

$$\blacksquare\ \text{expand}\left(\frac{2\cdot x^3+4\cdot x+10}{x^4+x^2+2\cdot x}\right) \qquad \frac{-2\cdot x}{x^2-x+2} + \frac{1}{x^2-x+2} - \frac{1}{x+\ldots}$$

```
...x^3+4x+10)/(x^4+x^2+2x))
```

$$\blacksquare \int \left(\frac{2\cdot x^3+4\cdot x+10}{x^4+x^2+2\cdot x}\right)dx \qquad -\ln\left(\frac{\left|(x+1)\cdot\left(x^2-x+2\right)\right|}{\left(|x|\right)^5}\right)$$

```
...^3+4x+10)/(x^4+x^2+2x),x...
```

The top display on the right shows how we can use the command **expand(** to express a quotient as a sum of partial fractions. The lower display shows the output from the integrate command. In exercise ❷ you will prove that this expression is equivalent to the expression derived above.

EXERCISES & INVESTIGATIONS

❶. Describe each integration technique and give an example of an integral which can be computed using that method.
a) integration by parts
b) integration by partial fractions
c) integration by trigonometric substitution

❷. Prove that the indefinite integral obtained by the TI-89 in *Worked Example* 3 is equivalent to the integral computed using partial fractions.

❸. Calculate $\int \sqrt{x}\sqrt{1-x}\,dx$ using the integral command (F3 2) from the home screen of your TI-89. Show that the indefinite integral you obtain on your display is the same as that given on page 70.

❹. Integrate by parts.

a) $\int xe^{x}dx$ b) $\int x\sqrt{x+3}\,dx$ c) $\int x^2\cos x\,dx$

❺. Integrate by trigonometric substitution.

a) $\int \sqrt{9-4x^2}\,dx$ b) $\int x^2\sqrt{4-x^2}\,dx$ c) $\int \left(\sqrt{16-x^2}\right)^{-\frac{3}{2}}dx$

❻. Integrate $\int \frac{5x^4+9x^3+7x^2+3x+2}{x^5+2x^4+2x^3+2x^2+x}\,dx$

by using the **expand (** command as in *Worked Example* 3 to express the integrand as a sum of partial fractions.

❼. Integrate by parts.

a) $\int e^{ax}\cos bx\,dx$ b) $\int x\ln x\,dx$

c) $\int \sin^2 x\,dx$ d) $\int \ln x\,dx$

❽. Integrate.

a) $\int x^3(1-x^2)^{\frac{1}{2}}\,dx$ b) $\int x(1+x)^{\frac{1}{2}}\,dx$ c) $\int \frac{x^2+4x-3}{x^3-2x^2-x+2}\,dx$

❾. Integrate.

a) $\int \frac{x}{x^2+x-6}\,dx$ b) $\int \frac{8-x}{x^3+4x}\,dx$ c) $\int \sin^3 x\,dx$

❿. Integrate.

a) $\int_0^3 \sqrt{9-x^2}\,dx$ b) $\int_{\sqrt{3}}^{2\sqrt{2}} \frac{1}{x\sqrt{x^2-2}}\,dx$

c) $\int_0^2 x^2\sqrt{4-x^2}\,dx$ d) $\int_0^{\pi/2} \sin^3 x\,dx$

⓫. Integrate $\int x^5\sqrt{1-x^3}\,dx$ using the substitution $z^2 = 1 - x^3$. Determine the area under the curve defined by $y = x^5\sqrt{1-x^3}$ between $x = 0$ and $x = 1$.

⓬. In the following investigation, you will discover how to deal with integrals of the form

$$\int \sin^n x\,dx \quad \text{and} \quad \int \cos^n x\,dx\ .$$

a) Expand the integral $\int \cos^2 x\,dx$ using integration by parts with $u(x) = \cos x$ and $dv(x) = \cos x\,dx$, so that $\int \cos^2 x\,dx$ and $-\int \cos^2 x\,dx$ appear on opposite sides of the equation. Transpose and solve for $\int \cos^2 x\,dx$.

b) Apply the procedure in part a) to the integral $\int \cos^4 x\,dx$ using the substitution $u(x) = \cos^3 x$, $dv(x) = \cos x\,dx$. Solve for $\int \cos^4 x\,dx$ and relate it to $\int \cos^2 x\,dx$.

c) Generalize the procedure in parts a) and b) to express $\int \cos^n x\,dx$ in terms of $\cos^{n-1} x \sin x$ and $\int \cos^{n-2} x\,dx$.

d) The formula you derived in part c) is called a *reduction formula* because it reduces the exponent n to $n-2$. Repeated reductions ultimately yield a form which is integrable.. Develop the analogous reduction formula for $\int \sin^n x\,dx$.

e) Use your reduction formulas developed above to evaluate

$$\int_0^{\pi/2} \cos^4 x\,dx \quad \text{and} \quad \int_0^{\pi/2} \sin^4 x\,dx$$

f) Use your TI-89 to check your answers in part e).

g) Derive formulas for

$$\int_0^{\pi/2} \sin^n x\,dx \quad \text{and} \quad \int_0^{\pi/2} \cos^n x\,dx$$

in the cases when (i) n is even (ii) n is odd. These are known as *Wallis's formulas.*

⓭. a) In investigation ❺ p. 19, we attempted to evaluate C(n), the area of the unit circle contained in the first quadrant. We used the numerical integration command **nInt (** to evaluate

$$\lim_{n\to\infty} \sum_{i=1}^{n} \sqrt{1-x_i^2}\,\Delta x_i$$

Find the indefinite integral of $\int \sqrt{1-x^2}\,dx$. Evaluate $\int_0^{\pi/2} \sqrt{1-x^2}\,dx$ and compare with your answer on p. 19.

Use your TI-89 to check both answers.

b) Integrate $\int \frac{b}{a}\sqrt{a^2-x^2}\,dx$. Evaluate $\int_0^{\pi/2} \frac{b}{a}\sqrt{a^2-x^2}\,dx$

Express the area enclosed by the ellipse $b^2x^2 + a^2y^2 = a^2b^2$ in terms of a and b.

Exploration 17 Applications of Integration in 2-Dimensions

How Far Did Mickey Mantle's Home Run Really Travel?

In *Exploration* 7 p. 34, we noted that the trajectory of the longest home run in major league history can be described by the equation $y = 0.9x - 0.0014x^2$. The variable, y denotes the height of the ball in feet when it has traveled x feet horizontally. But … how far has the ball *really* traveled???

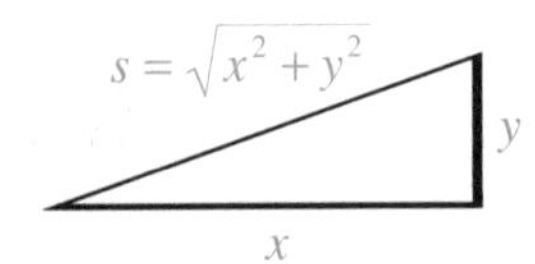

If the ball had traveled in a straight line, we would merely use the Theorem of Pythagoras to find the distance s. However, the ball traveled in a parabolic arc, so we need a way to calculate the *arc length* of the curve defined by $y = 0.9x - 0.0014x^2$. To devise a method for calculating arc length, we consider a very small interval over which the arc is almost linear. Then, for this small portion of the arc,

$$\Delta s \approx \sqrt{(\Delta x)^2 + (\Delta y)^2}$$

Dividing both sides of this expression by Δx yields $\dfrac{\Delta s}{\Delta x} \approx \sqrt{1 + \left(\dfrac{\Delta y}{\Delta x}\right)^2}$.

The derivative of the arc length is the limit of $\Delta s/\Delta x$ as $\Delta x \to 0$; that is,

$$\frac{ds}{dx} = \lim_{\Delta x \to 0} \frac{\Delta s}{\Delta x} = \lim_{\Delta x \to 0} \sqrt{1 + \left(\frac{\Delta y}{\Delta x}\right)^2} = \sqrt{1 + \left(\frac{dy}{dx}\right)^2}$$

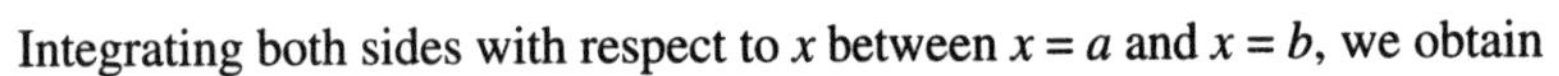

Integrating both sides with respect to x between $x = a$ and $x = b$, we obtain

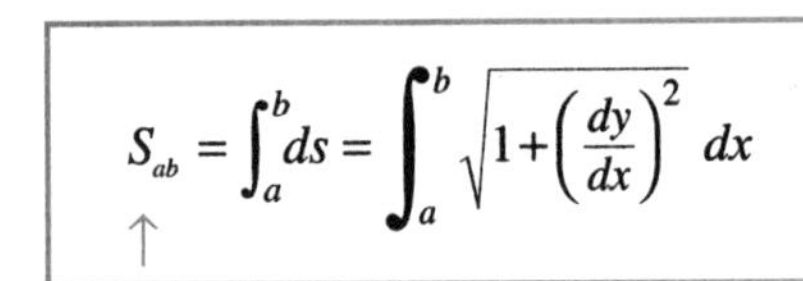

$$S_{ab} = \int_a^b ds = \int_a^b \sqrt{1 + \left(\frac{dy}{dx}\right)^2}\, dx$$

↑ arc length from a to b

Worked Example 1

a) Find an expression in x which gives the arc length of the curve $y = 0.9x - 0.0014x^2$, between 0 and x.

b) Use your expression to find the total distance traveled by the baseball on Mickey Mantle's famous home run.

c) Use your TI-89 to check your answers to parts a) and b).

Solution

a) To apply the formula for arc length given above, we calculate $dy/dx = 0.9 - 0.0028x$, so,

$$S = \int_0^x \sqrt{1 + (0.9 - 0.0028x)^2}\, dx = 0.0028 \int_0^x \sqrt{x^2 - 642.86x + 230867}\, dx \quad ①$$

In investigation ⑩ you will derive the formula

$$\int_0^x \sqrt{x^2 - bx + c}\, dx = \frac{(2x - b)\sqrt{x^2 - bx + c}}{4} + \frac{(4c - b^2)\ln\left|2\sqrt{x^2 - bx + c} + 2x - b\right|}{8} \quad ②$$

Substituting $b = 642.86$ and $c = 230867$ into ②, and then inserting this value into ① yields

$$S = 7.84 \times 10^{-6}(x - 321.43)\sqrt{x^2 - 642.86x + 230867} + 178.57 \ln\left|\sqrt{x^2 - 642.86x + 230867} + x - 321.43\right| \quad ③$$

SOLUTION (CONT'D)

b) To find the total distance traveled by the ball from its initial position at $x = 0$, to its termination at $x = 643$ feet, we substitute $x = 643$ into equation ③ and obtain 721.51 feet. That is, the ball traveled a distance of 721.51 feet along its arc length.

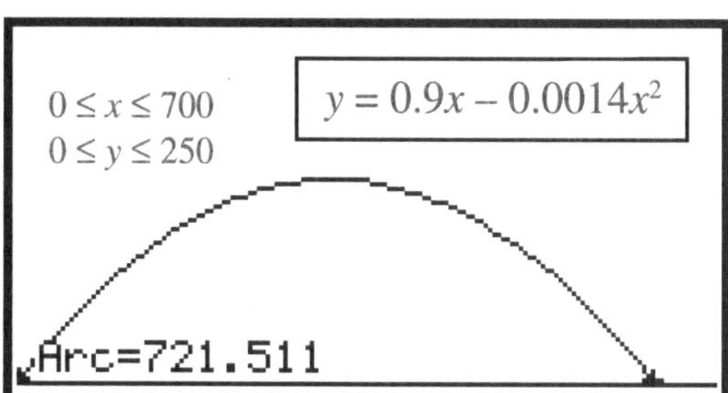

c) Performing the computations in parts a) and b) would have been extremely tedious without technology. To check that our answers are reasonable, we graph $y = 0.9x - 0\text{-}0014x^2$ in the window $0 \leq x \leq 700$; $0 \leq y \leq 250$. Then from the graph screen we press F5 alpha B and we enter 0 and 643 in response to the prompts first point? and second point?. The result is shown in the display on the right. It verifies the answer we obtained in part b).

```
■ ∫√(x² − b·x + c)dx
  (2·x − b)·√(x² − b·x + c)/4 − (b▸
∫(√(x^2-b*x+c),x)
```

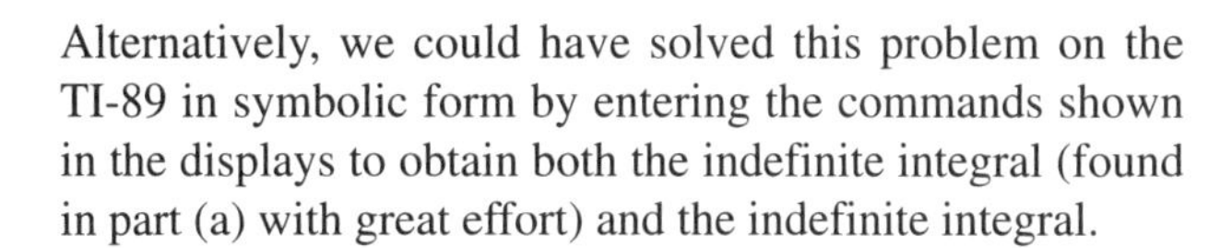

Alternatively, we could have solved this problem on the TI-89 in symbolic form by entering the commands shown in the displays to obtain both the indefinite integral (found in part (a) with great effort) and the indefinite integral.

```
■ ∫[0,643] (.0028·√(x² − 642.86·x▸
                              721.508
…028√(x^2-642.86x+230867)…
```

This example shows us that the use of technology in solving mathematics problems is essential if we wish to use numbers that describe real situations.

WORKED EXAMPLE 2

Graph the functions $y = \sin^2 x$ and $y = e^{-x}$ in the window $0 \leq x \leq \pi$; $0 \leq y \leq 1.2$.

Determine the points of intersection of these graphs and calculate the area between them.

SOLUTION

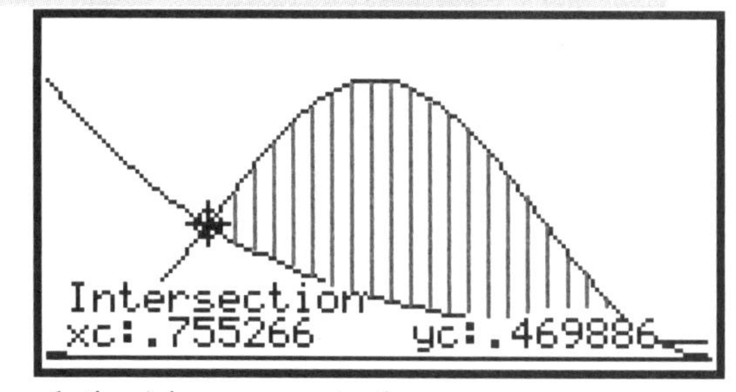

The display shows the graphs. To obtain the points of intersection, we press F5 5 from the graph screen. Then we press ENTER in response to each of the prompts 1st Curve? and 2nd Curve?. To obtain the coordinates of the left intersection point, we enter 0 and 1 (followed by ENTER each time) in response to the prompts Lower Bound? and Upper Bound? respectively. This yields the left intersection point (0.755…, 0.469…) shown in the display. Similarly, we find the right intersection point (2.905…, 0.054…).

The area bounded by the graphs between these two points is given by

$$\int_{0.755}^{2.905} \left(\sin^2 x - e^{-x}\right) dx = \left[\frac{1}{2}x - \frac{1}{2}\sin x \cos x + e^{-x}\right]\Bigg|_{0.755}^{2.905} = 1.0232\ldots \qquad \text{(See investigation ⑩)}$$

To evaluate this definite integral directly on the TI-89, we enter the commands and the intersection points on the command line as shown in the display. The output on the display confirms that the enclosed area is 1.0232…

```
■ ∫[.755,2.9] ((sin(x))² − e^-x)dx
                              1.0232
…(x)^2-e^(-x),x,.755,2.9)
```

EXERCISES & INVESTIGATIONS

❶. Explain how the arc length between two points A and B is different from the distance between those two points.

❷. a) Graph the function $y1(x) = 3x^3 - 7x^2 + 4x + 3$ in the region $-10 \le x \le 10; -10 \le y \le 10$.
b) Graph the tangent to $y1(x)$ at $x = 2$.
c) Express the area bounded by $y1(x)$, the tangent line and the axes as a difference of two definite integrals.
d) Evaluate these integrals by calculating the indefinite integrals and substituting the limits of integration.
e) Check your answer to part d) on your TI-89 using F5 alpha A and F5 7 from your graph screen.

❸. a) Graph the curves defined by $y1(x) = 1/x$ and $y2(x) = 8\cos x$ in the interval $0 \le x \le 2$.
b) Determine the points of intersection of these two graphs in this interval.
c) Calculate the area enclosed by these two curves between these two points of intersection.
d) Verify your answer on your TI-89 by using F5 7 from the graph screen.

❹. Determine the length of the curve defined by $y = 2x^{3/2}$ between $x = 0$ and $x = 7$.

❺. a) Write a definite integral which gives the length of the arc on a unit circle joining the points

$$\left(\frac{1}{\sqrt{2}}, \frac{1}{\sqrt{2}}\right) \text{ and } \left(\frac{\sqrt{3}}{2}, \frac{1}{2}\right)$$

b) Evaluate the definite integral and verify your answer by using the formula for the circumference of a circle.

❻. a) Graph the functions

$$y1(x) = \frac{2x}{\sqrt{3x^2 - 24x + 50}} \text{ and } y2(x) = \frac{8}{\sqrt{3x^2 - 24x + 50}}$$

in the window $0 \le x \le 10; 0 \le y \le 10$.
b) Determine a point of intersection of these two graphs that falls within this window.
c) Express the area of the region bounded by these graphs and the y-axis as a difference of definite integrals.
d) Evaluate the definite integrals in part c) by computing the indefinite integral of $y2(x) - y1(x)$.
e) Use your TI-89 to check your answer in part d).

❼. a) Prove that $\int \sec\theta\, d\theta = \ln|\sec\theta + \tan\theta| + C$.

Hint: Multiply the integrand by $\left[\dfrac{\sec\theta + \tan\theta}{\sec\theta + \tan\theta}\right]$ and then show that the numerator is the derivative of the denominator.

b) Use your formula to find the area between $\theta = 0$ and $\theta = \pi/4$ which lies below $y = \sec\theta$ and above $y = 1/2$.

❽. a) Graph the parabolic arch with equation $y = -4x^2 + 10$.
b) Calculate the length of this arch between $x = -2.5$ and $x = 2.5$.

Hint: Use the substitution $8x = \tan\theta$, and apply the formula
$$\int \sec^3\theta\, d\theta = \frac{1}{2}\left[\sec\theta\tan\theta + \ln|\sec\theta + \tan\theta|\right] + C$$

c) Use your TI-89 to verify your answer in part b).

❾. a) Graph the equation $y = x + \frac{1}{x}$.
b) Write a definite integral which gives the area under the graph between $x = \alpha$ and $x = \alpha + 1$ where $\alpha \ge 0$.
c) Compute the integral in part b) and graph the area as a function of α.
d) Find the value of α for which the area between $x = \alpha$ and $x = \alpha + 1$ is a minimum.
e) Show how you can obtain the result in d) without integrating the function $y = x + \frac{1}{x}$, by applying the Fundamental Theorem of Calculus.

❿. In this investigation, you will prove $\int_0^x \sqrt{x^2 - bx + c}\, dx$

$$= \frac{(2x-b)\sqrt{x^2 - bx + c}}{4} + \frac{(4c - b^2)\ln\left|2\sqrt{x^2 - bx + c} + 2x - b\right|}{8}$$

a) Set $z^2 - a^2 = x^2 - bx + c$, and calculate dz in terms of dx. (Note: z is a variable and a is a constant.)
b) Express the integrand in terms of z and a only.
c) To simplify the expression in b), substitute $z = a\sec\theta$, calculate dz and apply the identity $\sec^2\theta - 1 = \tan^2\theta$.
d) Integrate the expression in part c) using the formulas, for $\int \sec\theta\, d\theta$ and $\int \sec^3\theta\, d\theta$ given in exercises ❼ and ❽.
e) Substitute for θ to obtain the formula in x given above.

⓫. a) Describe the asymptote to the curve $x = \dfrac{ay^2}{a^2 + y^2}$.
b) Write an expression for a rectangular approximation to the area between the curve and its asymptote for $x \ge 0$.

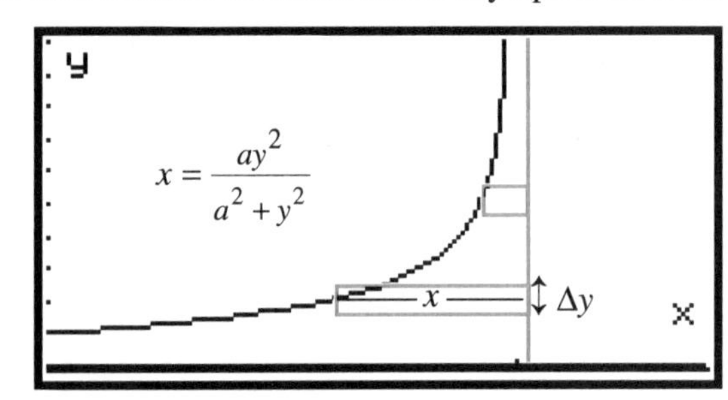

c) Write the limit of your expression in part b) as an integral.
d) Evaluate the integral in part c) to determine the area between the curve and its asymptote.

EXPLORATION 18 APPLICATIONS OF INTEGRATION IN 3-DIMENSIONS

CAN YOU PAINT AN INFINITE SURFACE WITH A POT OF PAINT?

In her computer modeling course, Karen has constructed the surface obtained by rotating the graph of $y = 1/x$ for $y \geq 1$ about the y-axis. How can she calculate the volume and surface area of her model?

WORKED EXAMPLE 1

a) Calculate the volume of the "solid" generated by rotating about the y-axis, the part of the graph defined by $y = 1/x$ and between $y = 1$ and $y = b$ (where $b > 1$). What is the limiting volume of the solid as $b \rightarrow \infty$?

b) Calculate the surface area of this solid and determine the limiting surface area as $b \rightarrow \infty$.

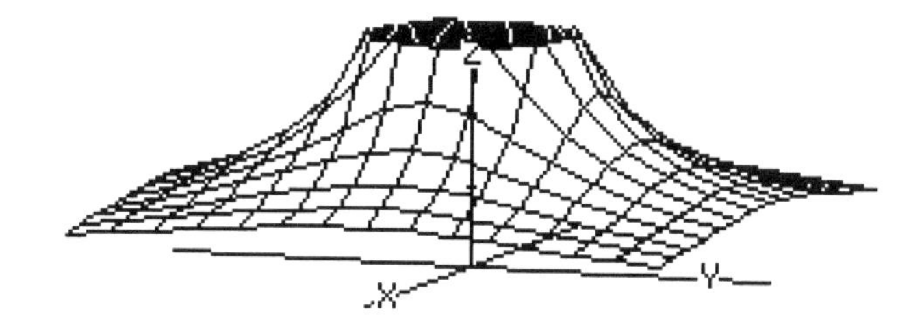

SOLUTION

In what follows, we develop the two standard techniques which are used to calculate volumes of solids of revolution.

Graph of $y = 1/x$ rotated about the y-axis

a) **The Disk Method**

The region under the curve $y = 1/x$ and between $y = 1$ and $y = b$ can be partitioned into a stack of n small rectangles of width x_i and height Δy. As the curve is rotated about the y-axis, each rectangle sweeps out a disk of radius x_i units and thickness Δy, as shown in the diagram. The volume of this disk is ΔV_i, where

$$\Delta V_i = \pi x_i^2 \Delta y$$

The total volume V swept out by the full stack of disks is

$$V = \lim_{n\to\infty} \Delta V_i = \lim_{n\to\infty} \sum_{i=1}^{n} \pi x_i^2 \Delta y$$

The limiting value of this sum is defined to be the integral $\int_1^b \pi x^2 dy$. Therefore,

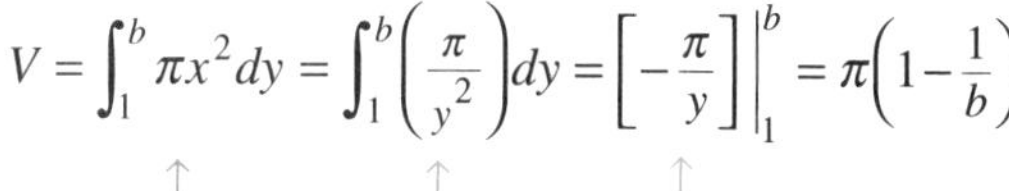

$$V = \int_1^b \pi x^2 dy = \int_1^b \left(\frac{\pi}{y^2}\right) dy = \left[-\frac{\pi}{y}\right]\Bigg|_1^b = \pi\left(1-\frac{1}{b}\right)$$

Substituting $x = 1/y$ Using the Power Rule of Integration

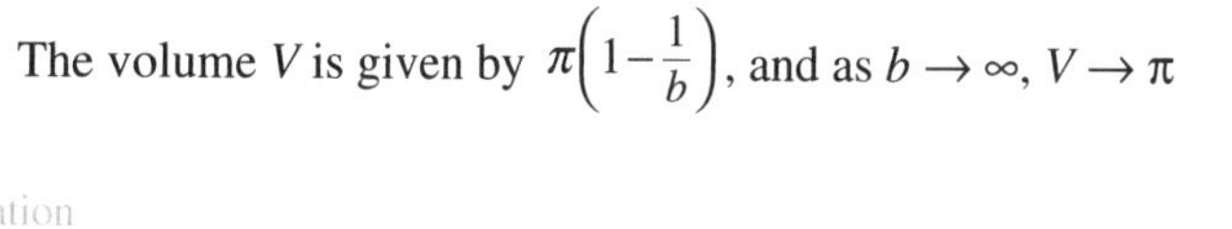

The volume V is given by $\pi\left(1-\frac{1}{b}\right)$, and as $b \rightarrow \infty$, $V \rightarrow \pi$

To calculate the surface area A, we proceed as above, however, we calculate ΔA the area of the surface of the disk. ΔA is the circumference of the disk $\times$ its slant height Δs. That is, $\Delta A = 2\pi x \Delta s$, and so,

Note: To get surface area we multiply by the element of arc length *not* thickness. Why?

$$A = \lim_{n\to\infty} \sum_{i=1}^{n} 2\pi x_i \Delta s = \int_{y=1}^{y=b} 2\pi x ds = 2\pi \int_{x=1/b}^{x=1} x\sqrt{1+\left(\frac{dy}{dx}\right)^2} dx$$

Substituting $\frac{dy}{dx} = -\frac{1}{x^2}$ yields

$$A = 2\pi \int_{1/b}^{1} x\sqrt{1+\left(\frac{-1}{x^2}\right)^2} dx = 2\pi \int_{1/b}^{1} \frac{\sqrt{x^4+1}}{x} dx$$

In investigation ❿, you will show that

$$\int_{1/b}^{1} \frac{\sqrt{x^4+1}}{x} dx = \left[\frac{2\sqrt{x^4+1} + 2\ln\left(\sqrt{x^4+1}-1\right) - 4\ln|x|}{4}\right]\Bigg|_{1/b}^{1}$$

As $b \rightarrow \infty$, this expression becomes infinite. That is, the surface area A is infinite although the volume V is π units!

SOLUTION (CONT'D)

a) **The Shell Method**

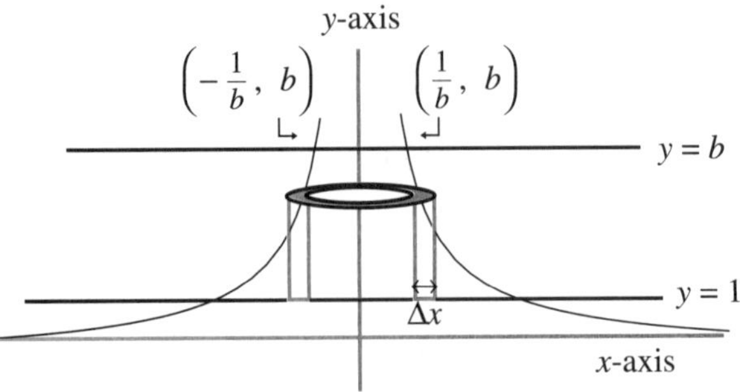

Alternatively, we can calculate the volume of a solid of revolution by considering the volume swept out by the n vertical rectangles that stretch from the line $y = 1$ to the curve $y = 1/x$. Each such rectangle has height $1/x_i - 1$ and width Δx. As it rotates about the y-axis, the rectangle sweeps out a cylindrical shell of radius x units, height $1/x_i - 1$ units and thickness Δx units. The volume of this shell is the difference in volumes of the outer and inner cylinders; i.e.,

$$\Delta V_i = \text{Volume of outside cylinder} - \text{Volume of inside cylinder}$$

$$= \pi(x_i + \Delta x)^2\left(\frac{1}{x_i} - 1\right) \quad - \quad \pi x_i^2\left(\frac{1}{x_i} - 1\right)$$

$$= \pi\left(\frac{1}{x_i} - 1\right)\left[2x_i\Delta x + (\Delta x)^2\right]$$

Therefore, the volume of revolution of the part of the curve $y = 1/x$ *between* $x = 1/b$ and $x = 1$ is given by V_{Outer} where

$$V_{Outer} = \lim_{n\to\infty}\sum_{i=1}^{n}\Delta V_i = \lim_{n\to\infty}\sum_{i=1}^{n}\pi\left(\frac{1}{x_i} - 1\right)\left[2x_i\Delta x + (\Delta x)^2\right] = \int_{x=1/b}^{x=1} 2\pi x\left(\frac{1}{x} - 1\right)dx$$

$$= 2\pi\int_{1/b}^{1}(1 - x)dx = 2\pi\left[x - \frac{x^2}{2}\right]\Bigg|_{1/b}^{1} = \pi - \frac{2\pi}{b} + \frac{\pi}{b^2}$$

← This is the volume of revolution of the part of the curve $y = 1/x$ *between* $x = 1/b$ and $x = 1$.

It remains only to calculate the volume of revolution of the part of curve $y = 1/x$ in the interval $-1/b \le x \le 1/b$ and below $y = 1$. This is merely a cylinder of radius $1/b$ and height $b - 1$. Its volume is

$$V_{Inner} = \pi\left(\frac{1}{b}\right)^2(b - 1)$$

The combined volume of revolution is

$$V = V_{Outer} + V_{Inner} = \pi - \frac{2\pi}{b} + \frac{\pi}{b^2} + \pi\left(\frac{1}{b}\right)^2(b-1) = \pi\left(1 - \frac{1}{b}\right)$$

As expected, the shell method yields the same result as the disk method. Although the shell method was a little more cumbersome for this solid of revolution, it is the more convenient method in many other cases.

b) On page 77, we computed the surface area by summing the surface areas of all the disks as x increases from $1/b$ to b. In investigation ❾, you will compute the surface area by summing all the disks as y increases from 1 to b, and verify that both methods yield the same surface area. As $b \to \infty$, the surface area becomes infinite.

We see that the volume of the solid of revolution generated by $y = 1/x$ is π cubic units yet its surface area is infinite. Suppose the surface of the solid were made of a thin porous substance. We could fill the inside of the solid of revolution with π cubic units of paint. The paint would fill the solid, and hence the entire inside surface of the solid. Since the surface is porous, the paint would seep through and cover the entire outer surface. That is, a finite amount of paint would cover an infinite surface. How can this be??

WORKED EXAMPLES

WORKED EXAMPLE 2

Determine the volume of the section of the paraboloid defined by

$$z = \frac{x^2}{9} + \frac{y^2}{16}$$

which is sliced off by the horizontal plane $z = 7$.

Paraboloid $z = \frac{x^2}{9} + \frac{y^2}{16}$

Plane $z = 7$

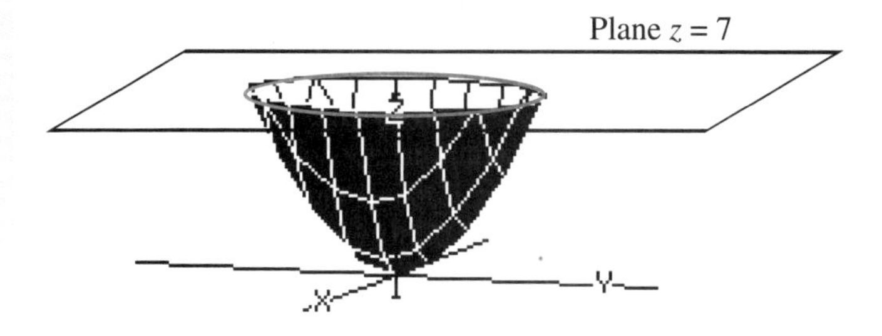

SOLUTION

The intersection of the plane z_i units above the x-y plane is the ellipse with equation

$$\frac{x^2}{9} + \frac{y^2}{16} = z_i$$

Recall: The area of the ellipse $\frac{x^2}{a^2} + \frac{y^2}{b^2} = 1$ is πab.

As in *Worked Example* 1, we can compute the volume of this solid by cutting the paraboloid (parallel to the x-y plane) into n elliptical slices of thickness Δz and computing the volume ΔV_i of an arbitrary slice. Then we take the sum of all ΔV_i over all slices as $n \to \infty$ to get the total volume V.

Area of ellipse $A_i = 12\pi z_i$

$A_i = 12\pi z_i$ ↕ Δz

Volume of elliptical disk is $\Delta V_i = 12\pi z_i \Delta z$

If A_i denotes the area of the elliptical slice with equation $\frac{x^2}{9} + \frac{y^2}{16} = z_i$ then $A_i = 12\pi z_i$

The volume of the slice at height z_i is given by $\Delta V_i = (12\pi z_i)\Delta z$. Hence,

$$V = \lim_{n\to\infty} \sum_{i=1}^{n} \Delta V_i = \lim_{n\to\infty} \sum_{i=1}^{n} 12\pi z_i \, \Delta z = \int_0^7 12\pi z dz = \left[12\pi \frac{z^2}{2}\right]\Bigg|_0^7 \text{ or } 294\pi.$$

That is, the volume of the paraboloid sliced off by the horizontal plane is 294π cubic units.

isualizing surfaces in 3-Dimensions

Before technology, there was little to help us visualize 3-D surfaces. Your TI-89 graphing calculator has a 3-D graphing mode which can help you develop your ability to think about solids and surfaces in three dimensions.

To access the 3-D graphing mode, press MODE and select 3D opposite Graph.

Then proceed as usual, by pressing ◆ [Y =] to define functions and pressing

◆ [WINDOW]

to set the window variables.

Try graphing the surface defined by

$$z = e^{-(x^2+y^2)}$$

in the window: $-2 \le x \le 2$
$-2 \le y \le 2$
$0 \le z \le 1$

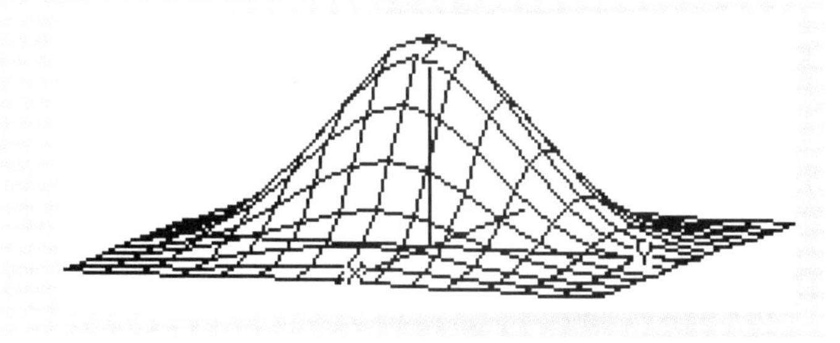

❶. In *Worked Example* 1 we discovered a surface which encloses a finite volume but has infinite surface area. We observed that such a surface could be filled with paint so that its inside, which is also infinite, is covered with paint. Explain why this is a paradox and determine whether there is a resolution to this paradox.

❷. a) Use the disk method in *Worked Example* 1 to find the volume of the solid generated by rotating about the y-axis, the part of the line $y = -x + 1$ in the first quadrant.

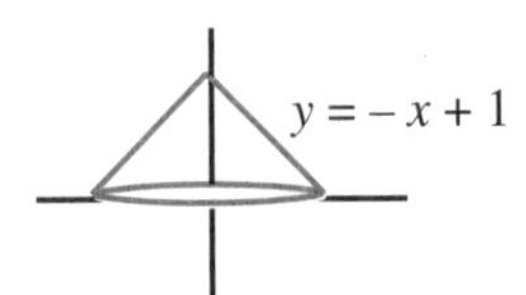

b) Write the equation of the surface generated in part a) where the z-axis is the vertical axis. Then graph this surface on your TI-89 using 3D mode to verify that your equation reduces to the given linear equation in the y-z plane.

c) Use the method of *Worked Example* 2 to determine the volume of the 3D solid and verify your answer to part a).

❸. a) Use the shell method to verify the volume of the solid of revolution that you computed in exercise ❷ a).
b) Calculate the surface area of the cone generated by the rotation of the line $y = -x + 1$ about the y-axis.

❹. a) Calculate the volume of the solid generated by rotating about the x-axis the region bounded by the parabola $y = 3\sqrt{x}$ the line $x = 8$ and the coordinate axes. Use the disk method.

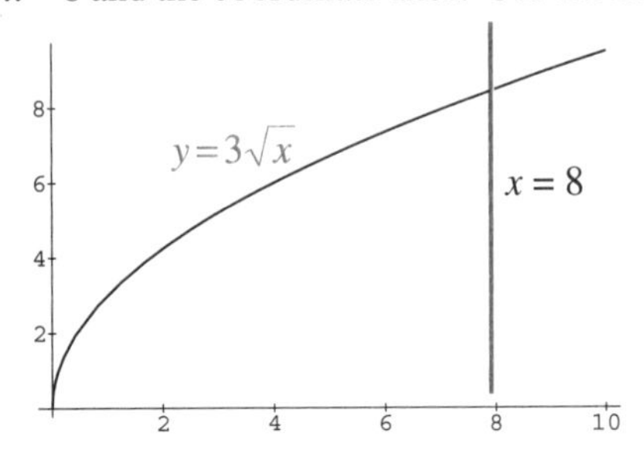

❺. Complete exercise ❹ using the shell method. That is, construct rectangular strips parallel to the x-axis and compute the volume of the cylindrical shells generated by the rotation. Check that your answer is the same as in exercise ❹.

❻. a) Graph the hyperboloid defined by the equation

$$z = \frac{1}{3}\sqrt{36 + 4x^2 + y^2}$$

in the window $-20 \le x \le 20$; $-40 \le y \le 40$; $0 \le z \le 10$.

b) Compute the volume of the hyperboloid below the plane defined by $z = 8$

❼. a) Suppose that a curve defined by $y = f(x)$ (where $f(x)$ is differentiable, $f'(x)$ is continuous and $f(x) \ge 0$ on $[a, b]$) is rotated about the x-axis. Prove that the surface area S of the solid of revolution between $x = a$ and $x = b$ is given by

$$S = 2\pi\int_a^b f(x)\sqrt{1 + f'(x)^2}\,dx$$

b) Use the formula you proved in part a) to derive the formula for the surface area of a sphere of radius R.

❽. If $f(x)$ is a continuous function and on $f(x) \ge 0$ on $[a, b]$, prove that the volume of the solid of revolution between $x = a$ and $x = b$ generated by rotating $f(x)$ about the x-axis is given by

$$V = \pi\int_a^b f(x)^2\,dx$$

❾. a) On page 74, we showed that the arc length S_{ab} is

$$S_{ab} = \int_a^b \sqrt{1 + \left(\frac{dy}{dx}\right)^2}\,dx \quad \text{Prove } S_{ab} = \int_{y(a)}^{y(b)} \sqrt{1 + \left(\frac{dx}{dy}\right)^2}\,dy$$

b) Use the result in part a) to calculate the surface area in *Worked Example* 1 by integrating with respect to y.

❿. In this investigation you will derive the following equation which we needed in *Worked Example* 1.

$$\int_{1/b}^{1} \frac{\sqrt{x^4+1}}{x}\,dx = \left[\frac{2\sqrt{x^4+1} + 2\ln\left(\sqrt{x^4+1} - 1\right) - 4\ln|x|}{4}\right]_{1/b}^{1}$$

a) To simplify the integrand of $\int_{1/b}^{1} \frac{\sqrt{x^4+1}}{x}\,dx$ substitute $u = \sqrt{x^4+1}$ and calculate du.

b) After the integrand is expressed as a function of u, break it into partial fractions and integrate.

c) Convert your expression in u to an expression in x. Then prove that the expression you have is equivalent to the expression on the right side of the above equation.

⓫. a) Use your TI-89 in 3D mode to graph the surface defined by

$$z = \frac{1}{\sqrt{x^2 + y^2}} \qquad \begin{array}{c} -5 \le x \le 5 \\ -5 \le y \le 5 \\ 0.3 \le z \le 0.8 \end{array}$$

b) Calculate the volume contained within this surface for $a \le z \le b$ where a, b are both positive numbers.

c) Use your result in part b) to determine the volume contained within this surface as $b \to \infty$.

d) Calculate the surface area in the interval $a \le z \le b$.

e) Write the equation of a curve in two dimensions which would generate this surface if rotated about the z-axis.

SELECTED SOLUTIONS

Exploration 1

1. a) A *finite sequence* is a function $u(n)$ defined on the set of positive integers from 1 to N, such that $u(n)$ is the nth term of the sequence. 2, 4, 6, …30 is a finite sequence having 15 terms.

b) An *infinite sequence* is a function $u(n)$ defined on the set of all positive integers, such that $u(n)$ is the nth term of the sequence. The sequence 2, 4, 6, 8, …, $2n$, … is an infinite sequence.

c) An arithmetic sequence is any sequence for which the nth term can be expressed in the form $u(n) = a + (n-1)d$ where a and d are constants. The first term is a and the common difference is d.

d) A geometric sequence is any sequence for which the nth term can be expressed in the form $u(n) = ar^{n-1}$ where a and r are constants. The first term is a and the common ratio is r.

2. The expression "$\lim_{n\to\infty} u(n)=L$" means that as n increases without limit, the terms of the sequence $u(n)$ approach arbitrarily closely to the value L, so that they are eventually larger than any given number less than L and smaller than any number exceeding L.

3. The statement is true by definition.

4. a) arithmetic sequence {95, 86, 77, 68, 59, 50, 41, 32, 23, 14}
b) geometric sequence {1, 3, 9, 27, 81, 243, 729, 2187, 6561, 19683}
c) neither type {6, 4, 0, –8, –24, –56, –120, –248, –504, –1016}
d) geometric sequence
{1/2, 1/4, 1/8, 1/16, 1/32, 1/64, 1/128, 1/256, 1/512, 1/1024}
e) neither type
{3, 4, 4.63, 5.063, 5.378, 5.619, 5.808, 5.96, 6.086, 6.192}

5. a) $u1(18) = -58$ b) $u2(18) = 129\ 140\ 163$
c) $u3(18) = -262\ 136$ d) $u1(18) = 1/(262\ 144)$
e) $u5(18) = 6.66246\ldots$

6. a) & b) Tracing reveals that $u4(20) = 0.00000095367$ and $u5(20) = 6.7275$.

Hint: You can trace directly to the point with $n = 20$, by pressing F3, entering 20 and pressing ENTER.

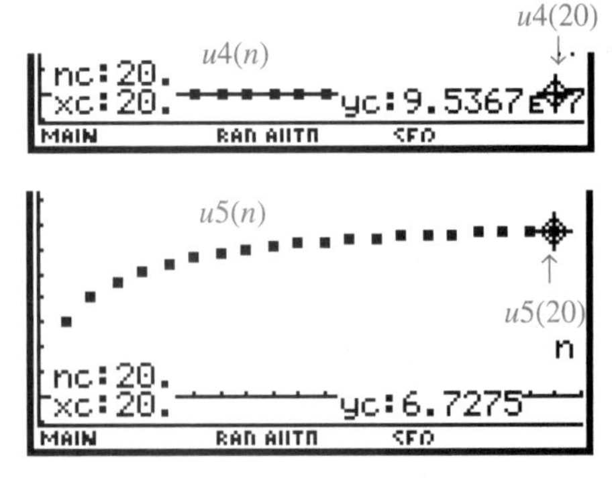

c) $\lim_{n\to\infty} u4(n) = 0$ and $\lim_{n\to\infty} u5(n) = e^2 \approx 7.38906$.
This converges very slowly to 0.

8. a) $\lim_{n\to\infty} u(n) = 1$ b) $\lim_{n\to\infty} u(n) = 0$

c) $\lim_{n\to\infty} u(n) = 0.739\ldots$ d) $\lim_{n\to\infty} u(n) = 0.642857\ldots$

Exploration 1

11. a) The displays show the sequences corresponding to $\lambda = 1.6$, 3.1, and 3.5 respectively and their graphs.

```
 u1=1.6·u1(n - 1)·(1 - u1(n - 1))
ui1=.6
 u2=3.1·u2(n - 1)·(1 - u2(n - 1))
ui2=.6
 u3=3.5·u3(n - 1)·(1 - u3(n - 1))
```

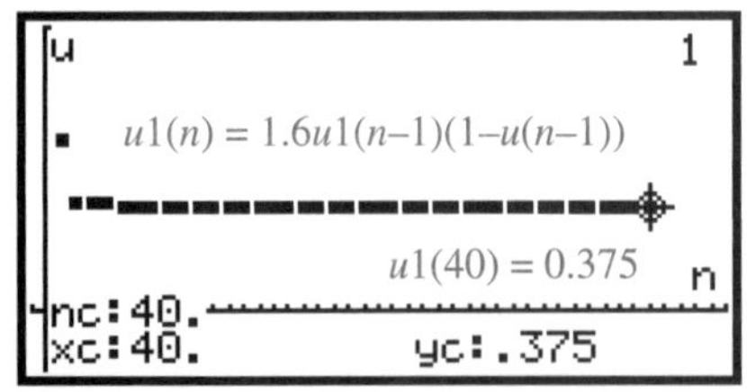

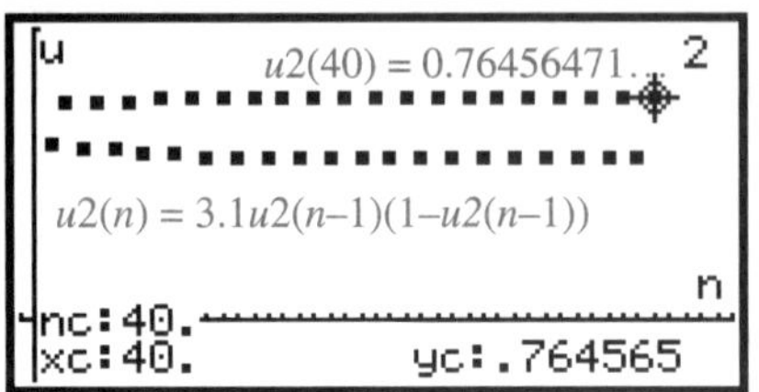

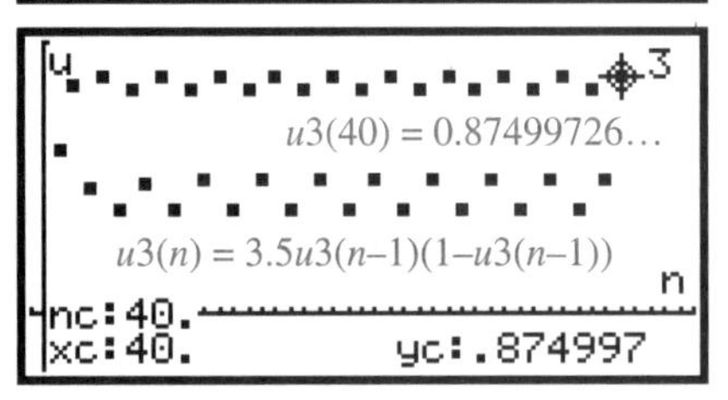

Hint: You can trace directly to the point with $n = 40$, by pressing F3, entering 40 and pressing ENTER.

b) & c) The web plots and the stable points are shown below.

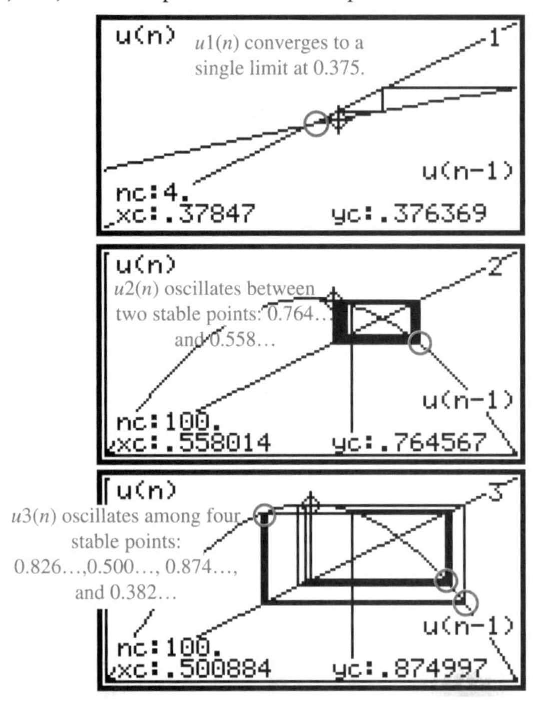

Selected Solutions to Exercises & Hints for the Investigations

Exploration 2

1. It is true that the runner has an infinite number of finite segments to traverse, and that each segment will require a finite time. However the assumption that an infinite number of these finite times must have an infinite total is not valid. We saw on page 12, that it is possible for an infinite number of terms in a series to sum to a finite number.

2. $\Sigma\, u(n)$ is the sum of the quantities $u(n)$ as the variable, n assumes in turn all positive integral values between and including a (the initial value of n) and b (the final value of n).

3. a) $1 + 8 + 27$ and 8000 b) $1/2 + 1/4 + 1/8$ and $1/2^{100}$
c) $1 + 1/4 + 1/9$ and $1/400$ d) $5 + 14 + 29$ and 1202

4. a) $\sum_{n=0}^{60} 2^{-n} = 2$ b) $\sum_{n=1}^{100} n^{-1} = 5.18737\ldots$

c) $\sum_{n=0}^{100} (2n+1) = 10\,201$ d) $\sum_{n=0}^{7} (-1)^n 2^{2n+1} = -26\,214$

Remember: To convert from fraction to decimal form, press: ◆ ENTER

5. a) 1300 b) 0.700653… c) 1.201860…

6. a) $\sum_{k=0}^{n-1} (a+kd)$ b) Sum $= \frac{n}{2}[2a+(n-1)d]$ c) 1683

7. a) $\frac{N(N+1)}{2}$ b) $\frac{N(N+1)(2N+1)}{6}$ c) $\left(\frac{N(N+1)}{2}\right)^2$ d) $\frac{1-x^{n+1}}{1-x}$ if $x \neq 1$

8 e) (i) 1/4 (ii) 3/5 (iii) $66\frac{2}{3}$

9. d) The commands and the formula are shown in the display. A little simplication is required to put the formula in standard form.

$$\sum_{k=1}^{n}\left((a+(k-1)\cdot d)\cdot r^{k-1}\right)$$
$$\frac{((a+d\cdot(n-1))\cdot r-a-d\cdot n)\cdot}{(r-1)^2}$$

```
Σ((a+(k-1)d)*r^(k-1),k,1,...
```

$$S_n = \frac{a(r^n-1)}{r-1} + \frac{rd[1-nr^{n-1}+(n-1)r^n]}{(r-1)^2}$$

Is the Harmonic Series Finite?

When Gottfried Leibnitz co-inventor of calculus died in 1716, he went to his grave believing that the harmonic sum was finite. However, it was subsequently proved that the harmonic sequence has an infinite sum. For an excellent presentation of the classical proof of this, see *A Mathematical Mosaic: Patterns & Problem Solving* by Ravi Vakil (p. 80) available through this publisher.

a) 5.18737… b) 7.48547… c) 9.09450…

The commands are shown in the command line of this display.

$$\sum_{n=1}^{5000}\left(\frac{1}{n}\right) \qquad 9.09451$$

```
Σ(1/n,n,1,5000)
```

Exploration 3

1. Your explanation should include these elements.
   - We approximate the true area by filling it with n triangles or rectangles which cover most of the area.
   - We express the area $S(n)$ of these n rectangles as a function of n. $S(n)$ is a lower bound for the true area.
   - Then we cover the true area with n rectangles or triangles which form an upper bound $T(n)$ of the true area.
   - We evaluate the limits of $S(n)$ and $T(n)$ and if they are the same, then the true area is equal to that common limit.

2. a) The area of the large triangle is $\sin(\pi/n)\times\cos(\pi/n)$, or $1/2\sin(2\pi/n)$ so $P(n) = n/2 \times \sin(2\pi/n)$.
   b) The area of the large triangle is $\tan(\pi/n)$, so $Q(n) = n\tan(\pi/n)$
   c) Both limits are π as shown in the displays below.

$$\lim_{n\to\infty}\left(\frac{n}{2}\cdot\sin\left(\frac{2\cdot\pi}{n}\right)\right) \qquad \pi$$

```
limit(n/2*sin(2π/n),n,∞)
```

$$\lim_{n\to\infty}\left(n\cdot\tan\left(\frac{\pi}{n}\right)\right) \qquad \pi$$

```
limit(n*tan(π/n),n,∞)
```

   d) The true area of the circle of radius 1 is sandwiched between the upper and lower bounds which converge to π, so the area of the unit circle is π units. A circle of radius r is a magnification of the unit circle by a scale factor r, so its area is increased by the factor r^2. Therefore, a circle of radius r has area πr^2.

3. a) $T(5) = 2/5[f(0) + f(2/5) + f(4/5) + f(6/5) + f(8/5)]$
   b) As shown in the display, this area is $152/25 = 6.08$.

$$2/5\cdot\sum_{k=0}^{4}\left(-\left(\frac{2\cdot k}{5}\right)^2+4\right) \qquad \frac{152}{25}$$

```
2/5*Σ(-(2k/5)^2+4,k,0,4)
```

   c) $T(n)$ is shown in the display as a sum and as an expression.

$$\frac{2}{n}\cdot\sum_{k=0}^{n-1}\left(-\left(\frac{2\cdot k}{n}\right)^2+4\right) \qquad \frac{4\cdot(4\cdot n^2+3\cdot n-1)}{3\cdot n^2}$$

```
...n*Σ(-(2k/n)^2+4,k,0,n-1)
```

   d) The display shows that $T(n)$ approaches the same limit as $S(n)$, that is, 16/3.

$$\lim_{n\to\infty}\left(\frac{2}{n}\cdot\sum_{k=0}^{n-1}\left(-\left(\frac{2\cdot k}{n}\right)^2+4\right)\right) \qquad 16/3$$

```
...(2k/n)^2+4,k,0,n-1),n,∞)
```

4. The four numerical integrals are shown below.

```
nInt(x^2 - 4, x, -2, 2)        -10.66666667
nInt(log(x), x, 1, 3)          .5627748004
nInt(1/(x + 5), x, 0, 4)       .5877866649
nInt(1 + x^(1/3), x, 1, 8)     18.25
```

Exploration 3 (cont'd)

5. a) & b) $S(n) = abC(n)$ where the expression for $C(n)$ is shown in the command line.

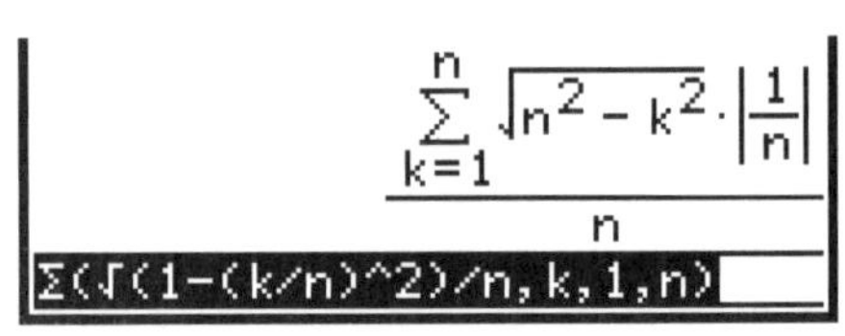

The fact that the display yields an expression in sigma notation, reveals that the TI-89 was not able to express the summation in closed form, i.e. as a function of n only. This suggests that the sum of these radicals is not representable as a simple algebraic expression in n.

c) When we use the limit function, and apply it as $n \to \infty$, we obtain the display shown here.

$$\lim_{n\to\infty}\left(\frac{\left|\frac{1}{n}\right| \cdot \sum_{k=1}^{n}\sqrt{n^2-k^2}}{n}\right)$$

```
...1-(k/n)^2)/n,k,1,n),n,∞)
```

We see that the TI-89 is unable to evaluate the limit of this series. This means that we must seek an alternative method of approximating $C(n)$.

d) We obtain the display below, confirming that the area of the unit circle in the first quadrant is $\pi/4$.

```
■ nInt(√(1 - x²), x, 0, 1)
                      .785398
nInt(√(1-x^2),x,0,1)
```

The numerical integration command provides an effective method of estimating areas when the rectangular approximations cannot be written as sums in closed form. This is more the rule than the exception. In the last unit of this book, you will study formal integration techniques that will broaden the range of curves for which you can represent contained areas in closed form.

Exploration 4

1. a) The nth partial sum of the series $a_1 + a_2 + a_3 + \ldots + a_k + \ldots$ is $S(n) = a_1 + a_2 + a_3 + \ldots + a_n$

b) If S denotes the sum for a given series, then $S = \lim_{n\to\infty} S(n)$.

2. a) As x approaches arbitrarily close to c, $f(x)$ approaches arbitrarily close to L.

b) If substitution of x_0 into the expression for $f(x)$, yields an undefined (i.e. meaningless expression) such as $\frac{0}{0}$ or $\frac{\infty}{\infty}$, then $f(x)$ is said to be *indeterminate*.

c) The numerator and denominator have a common factor that approaches zero as the independent variable approaches the limit. We must factor both and cancel out the common factor(s) and then evaluate the limit.

Exploration 4 (cont'd)

3. a) $\lim_{x\to\infty}\frac{f(x)}{g(x)} = \infty$ b) $\lim_{x\to\infty}\frac{f(x)}{g(x)} = 0$ c) $\lim_{x\to\infty}\frac{f(x)}{g(x)} = \frac{a_n}{b_m}$

4. a) i) $\frac{-4(3^9-1)}{2} = -39\,364$ ii) $\frac{7\left(1-\left(\frac{1}{4}\right)^{15}\right)}{\frac{3}{4}} = 9.3333$

iii) $\frac{5\left(1-\left(\frac{2}{3}\right)^{30}\right)}{\frac{1}{3}} = 14.9999$

5. a) i) $\frac{6}{\left(1-\frac{7}{8}\right)} = 48$ ii) $\frac{4}{\left(1+\frac{3}{4}\right)} = 2.28571\ldots$

b)

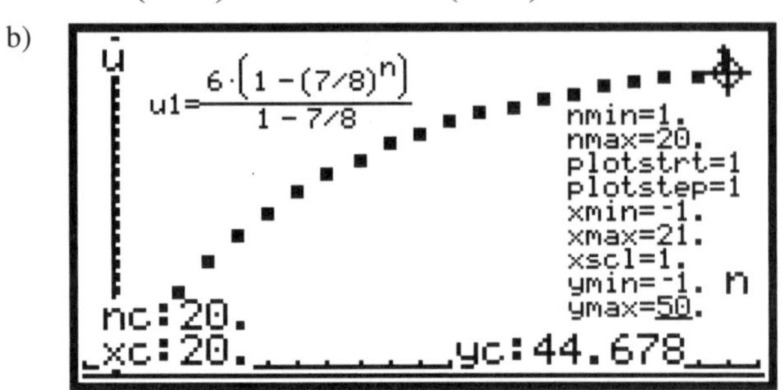

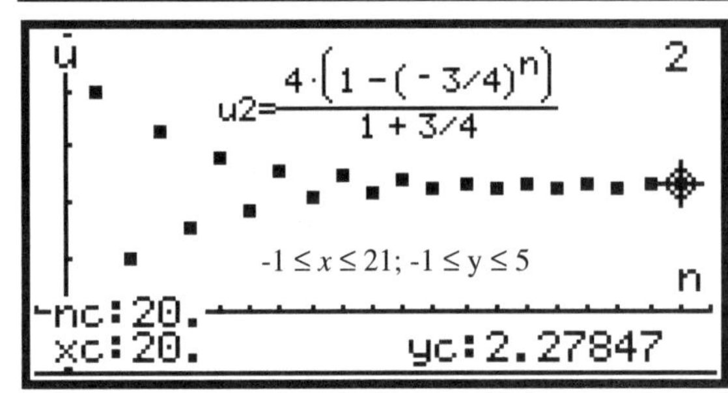

6. a) ∞ b) 54 c) 7/2 d) $-23/12$

7. a) -7 b) $\frac{-3\sqrt{2}-5}{2}$ c) 1/3 d) 3

8.

```
■ lim  (x²·(1 - cos(x⁻¹)))
  x→∞
                               1/2
■ lim  (x²·(1 - cos(x⁻¹)))      0
  x→0
...it(x^2(1-cos(x^-1)),x,0)
```

9. a) $\lim_{x\to 0}\frac{\sin kx}{x} = \lim_{x\to 0}\frac{k\sin kx}{kx} = k\lim_{kx\to 0}\frac{\sin kx}{kx} = k$

b) $\lim_{n\to\infty} 2n\sin\left(\frac{\pi}{n}\right) = 2\pi \lim_{\left(\frac{\pi}{n}\right)\to 0}\frac{\sin\left(\frac{\pi}{n}\right)}{\left(\frac{\pi}{n}\right)} = 2\pi$

10. a) No limit exists b) 0 c) $\frac{864}{x} + \frac{216}{x^2} + \frac{24}{x^3} + \frac{1}{x^4}$ d) 0

11. $\sqrt{30}/2 \approx 2.7386$ 12. e) (i) e^{ab} (ii) e^{ab}

Exploration 5

1. a) $v_{av} = \Delta s/\Delta t$

b) To determine the average velocity between time t_0 and t_1, locate the points on the distance-time graph corresponding to these times. Calculate the slope of the secant joining these two points.

2. a) $a_{av} = \Delta v/\Delta t$

b) To determine the average acceleration between time t_0 and t_1, locate the points on the velocity-time graph corresponding to these times. Calculate the slope of the secant joining these two points.

c) The average velocity is the arithmetic mean of the velocities at $t = 2.8$ and $t = 9.2$, which is $9.8[2.8 + 9.2]/2 = 58.8$ *m/s*.

3. Yes. The development in Worked Example 1 shows that if the acceleration is a constant a, then $v = v_0 + at$ where v_0 is the initial velocity. That is, the velocity is a linear function of time.

4. a) The cannon ball reaches the ground when $4.9t^2 = 56$; i.e. when $t = \sqrt{(56/4.9)} \approx 3.38$. Therefore it hits the ground in about 3.38 *s*. No. The cannon ball is accelerating, so it travels farther in the second half of the time interval. The distance traveled after 3.5 *s* would be less than half the distance traveled in that one-second interval.

5. We graph $y = -0.07x^3 + 3.15x^2 + 1.2x$ in the window: $0 \le x \le 15$ and $0 \le y \le 1000$.
a) The value command yields $y = 90.24$ and 156.65 *m* respectively for the times $t = 5.5$ and 7.5. The display shows the latter value.

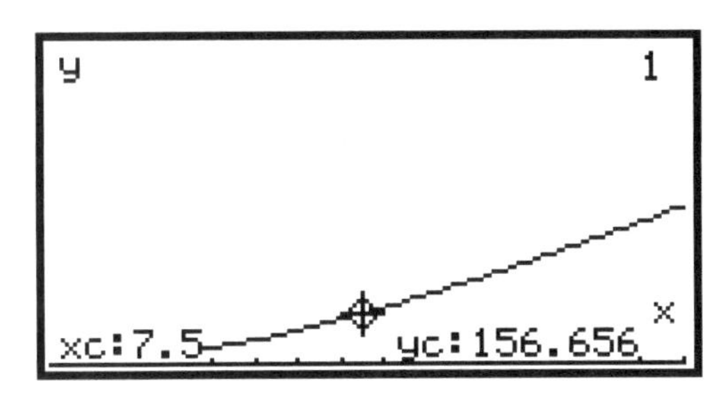

b) Average velocity is $(156.656 - 90.241)/2 \approx 33$ *m/s*.
c) Average velocity is $(93.210 - 90.241)/0.1 \approx 29.7$ *m/s*.
d) Average velocity is $(90.536 - 90.241)/0.01 \approx 29.5$ *m/s*. The instantaneous velocity at 5.5 *s* is close to 29.5 *m/s*.

6. The vertical velocity would be close to 27.4 *m/s* at $t = 7$.

7. a) $s + \Delta s = 4.9(t + \Delta t)^2$.
b) $\Delta s = 4.9(t + \Delta t)^2 - 4.9t = 9.8t(\Delta t) + (\Delta t)^2$.
d) $9.8t$

8. a) $2t$ b) $3t^2$

9. a) $y + \Delta t = -0.07(t + \Delta t)^3 + 3.15(t + \Delta t)^2 + 1.2(t + \Delta t)$.
d) $-0.21t^2 + 6.3t + 1.2$
e) 29.4975 *m/s*. We observe that this is very close to the average velocity computed in exercise 5 d).

Exploration 6

3a) & 4a) The graph of $y = 4x^2 - 7$ and its tangent line at $x = 3$.

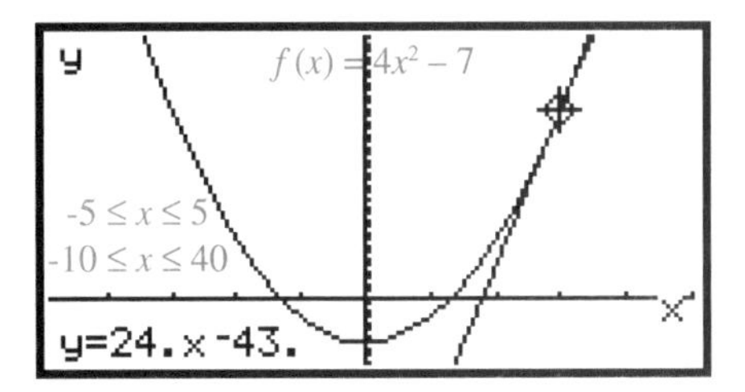

3b) & 4b) The graph of $y = 5x^3 + 6x$ and its tangent line at $x = 3$.

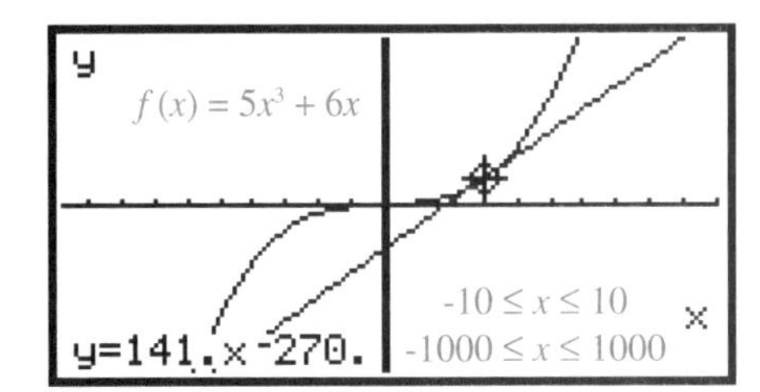

3c) & 4c) The graph of $y = 4x^3 + 3x^2 - 7$ and its tangent line at $x = 3$.

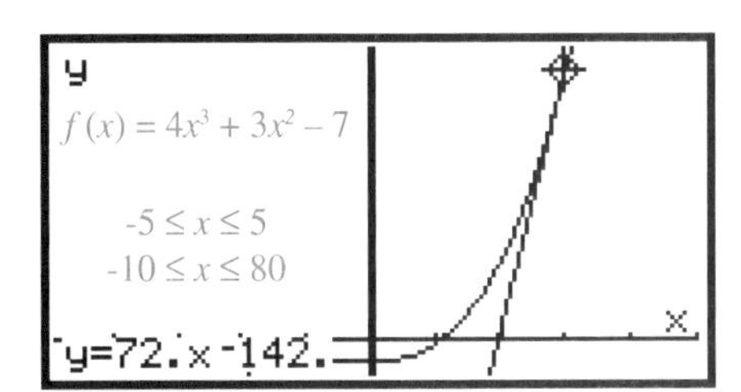

3d) & 4d) The graph of $y = 8x^3 + 4x - 9$ and its tangent line at $x = 3$.

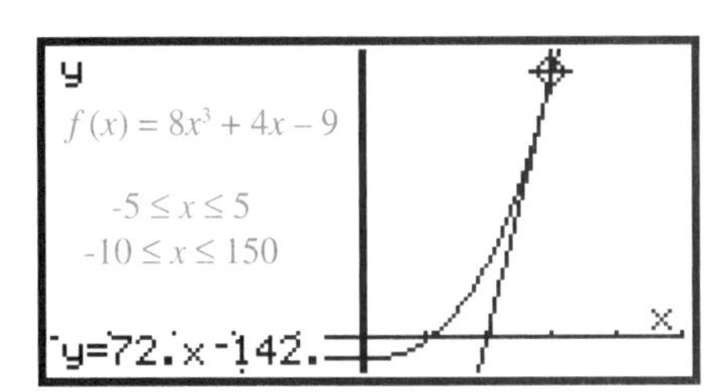

5. The velocity is given by $dy/dx = -0.21t^2 + 6.3t + 1.2$, so the velocity at $t = 5.5$ is 29.4975 *m/s*.

6. See the displays above in the answers to exercises 3 and 4.

7. b) We obtain the following display.

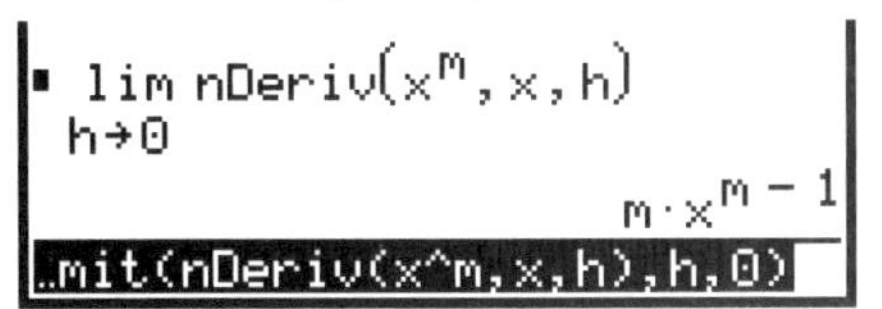

c) Substitution of a for x yields $f'(a) = ma^{m-1}$.

d) We obtain the display below.

$\frac{d}{dx}(x^m)$ $m \cdot x^{m-1}$

d(x^m,x)

Exploration 7

1. a) a parabola
b) 321.43 feet
c) slope of the secant 144.64/321.43 ≈ 0.45
d) The slope of the tangent to the trajectory is 0 when the ball is at its highest point, because the vertical velocity of the ball is momentarily zero as it ceases to travel upward and begins to travel downward. That is, $\Delta y = 0$ over the interval Δx, so $dy/dx = 0$.
e) This is a model which would apply perfectly if the atmosphere were thin. Air resistance and lift generated by a spin on the ball can alter the shape of this trajectory, but the model is pretty accurate.

2. a) The value $f(a)$ is said to be a *local minimum* of $f(x)$ if and only if $f(x) \geq f(a)$ for all values of x close to a.
b) The display shows that $y = x^3 + 8$ appears to flatten out near $x = 0$ but it has no local minimum because for $x < a, f(x) < f(a)$.

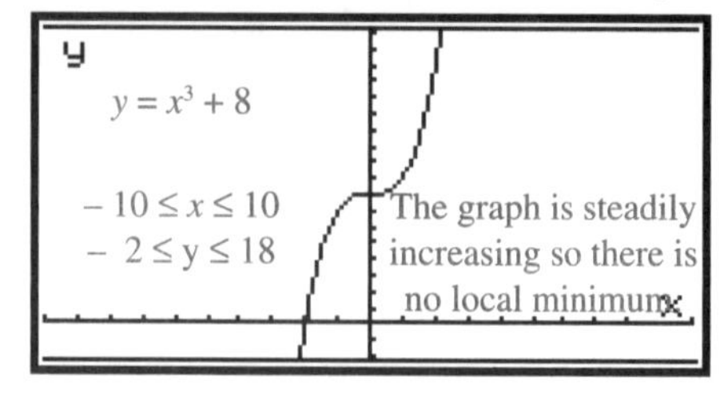

3. a) The display shows that there is a local maximum of 1.9065… at $x = -0.254…$

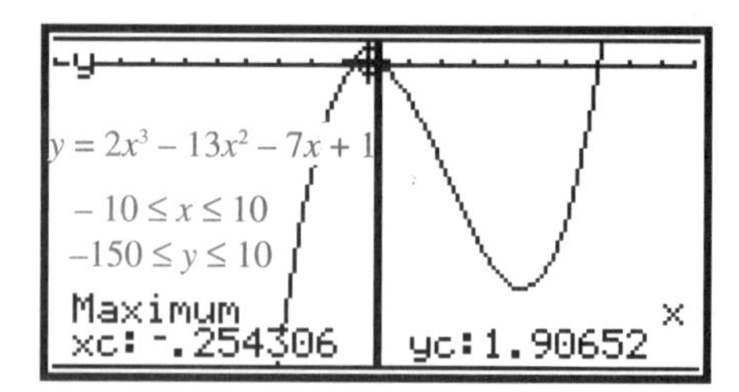

By repeating the process, we find a local minimum of –111.61… at $x = 4.587…$
b) The display shows there is a local maximum and a local minimum respectively at $x = (13 - \sqrt{211})/6$ and $x = (13 + \sqrt{211})/6$

```
■ solve(6·x^2 - 26·x - 7 = 0, x)
      √211 + 13          -(√211
  x = ---------  or  x = -------▸
          6                 6
solve(6x^2-26x-7=0,x)
```

When expressed in decimal form, the answers in a) & b) agree.

4. a) minimum at (7/6, 11/12)
b) minima at (1, –8) and (–2, –8); maximum at (–1/2, –47/16)
c) minimum at (1/2, –107/16)

5. a) None. The graph has constant slope and cannot change direction.
b) One. The graph is a parabola which opens upward or downward.
c) Two. The graph can cross the x-axis three times and change direction twice.
d) $n - 1$. Since $f'(x)$ is a polynomial of degree $n - 1$, the curve can change direction at most $n - 1$ times.

Exploration 7 (cont'd)

6. $f'(x)$ is to be a quadratic function with zeros at $x = 0$ and $x = 4$, such that $f'(x)$ is changing from positive to negative near $x = 0$ and from negative to positive near $x = 4$. One such function is $f'(x) = x(x - 4)$; that is $f'(x) = x^2 - 4x$. One such function with this property is $f(x) = x^3 - 6x^2$. This function is not unique since multiplication by a constant (a vertical stretch) and/or addition of a constant (vertical translation does not change the extrema.

7. Heron's formula states that the area A of a triangle is given by

$$A = \sqrt{s(s-a)(s-b)(s-c)}$$

where s is the semi-perimeter of the triangle. Since A is a maximum when A^2 is a maximum, it is sufficient to maximize A^2. The sides of the triangle are 5, 5 and x *cm* long, so $s = x/2 + 5$. We must maximize the function $f(x) = (x + 10)(x)(x)(10-x)/16$. That is, $f(x) = x^2(100-x^2)/16$. This is a maximum when $x = 5\sqrt{2}$. The triangle has maximum area when the angle between the equal sides is $\pi/2$.

8. We must minimize the distance from the point (x, y) to $(0, 0)$, or what is the same, the square of this distance, i.e. $x^2 + y^2$ for all points that satisfy the condition $y = 3x + 2$. The function,

$$f(x) = x^2 + (3x + 2)^2 \text{ or } f(x) = 10x^2 + 12x + 4$$

has a minimum of 74/5 at $x = -3/5$. The closest point to the origin is (–3/5, 1/5).

9. Proceeding as in exercise 8, we minimize the function $x^2 + y^2$, subject to the condition $y = x^2 - 4x + 3$. We find that the function $f(x) = x^4 - 8x^3 + 23x^2 - 24x + 9$ has a minimum at $x = 0.83…$ The y-coordinate of this point is $y = 0.38…$ The distance of this point from the origin is about 0.91…

10. a) Clearly the two lines are parallel with slope 2.7. The line through the origin which is perpendicular to these two lines has equation $y = -(2.7)^{-1}x$. We graph these three lines and select 5 from the F5 menu to get the points of intersection as shown in the display.

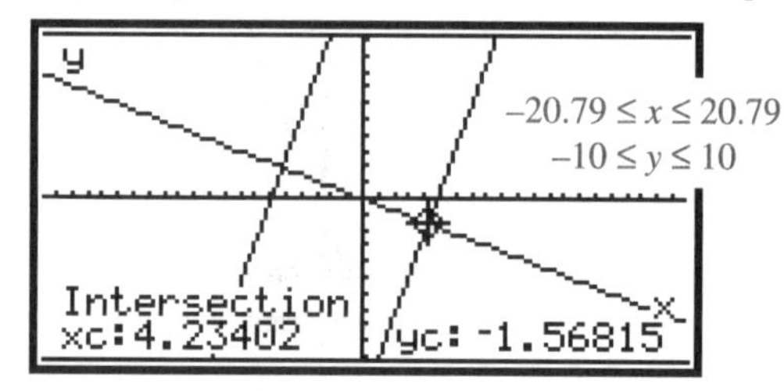

The points of intersection are (–5.21, 1.93) and (4.23, –1.56). The distance between these two points is 10.07 (to 2 decimal places). You can find this distance directly on your TI-89 by pressing F5 followed by 9 on your graphing screen and then entering the x coordinates of the points of intersection on the line $y = -(2.7)^{-1}x$.

11. The ships reach their closest proximity at 0.94 hours or 56.4 minutes after 7:00 a.m. At that time they are 52.68 km apart.

12. The profit function to be maximized is revenue minus cost, i.e. $f(x) = -0.35x^2 + 57x - 20$. This function achieves its maximum when the daily production is about 81 grummets and this yields a profit of about $2300.

13. The area function is $y = x(35 - 2x)$. The area is maximized when the length is 17.5 meters and the width is 8.75 meters.

Selected Solutions to Exercises & Hints for the Investigations

Exploration 8

1. A constant function, $f(x) = a$ is a line parallel to the x-axis. It has a slope of 0 for all x, so its first derivative is everywhere equal to 0.

2. This follows directly from the fact that multiplication of a function by a constant k multiplies its limit by a factor of k.

3. The graphs of $f(x)$ and $f'(x)$ are shown in the displays below.

3 a)

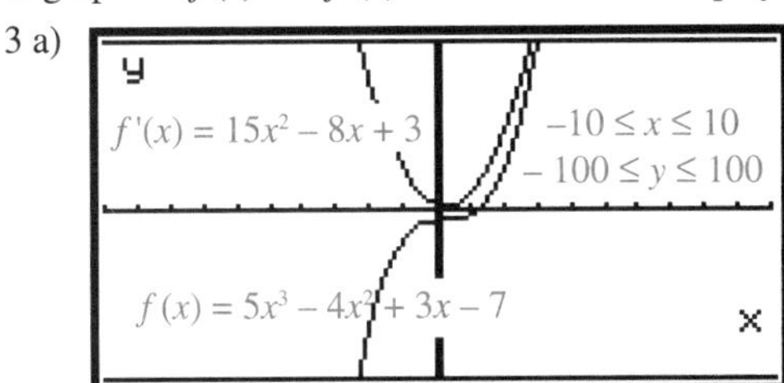

3 b)

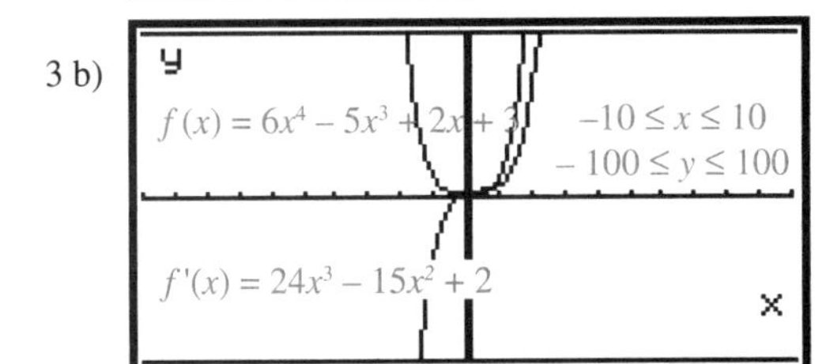

3 c)

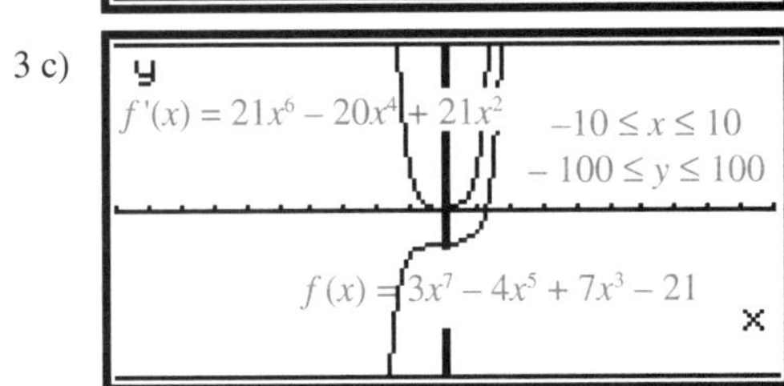

4. The displays show the graph, the tangent line and its equation.

4 a)

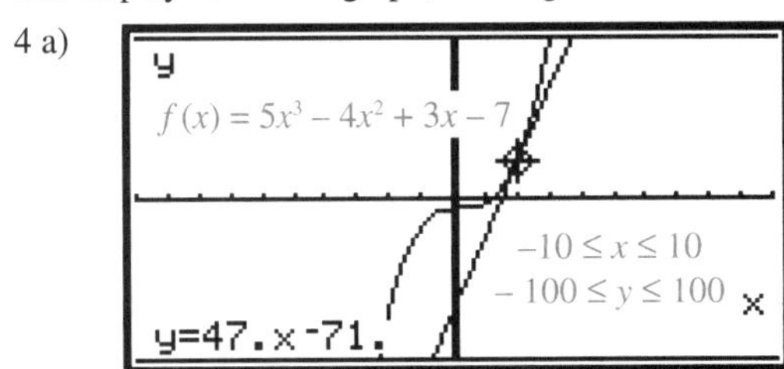

4 b)

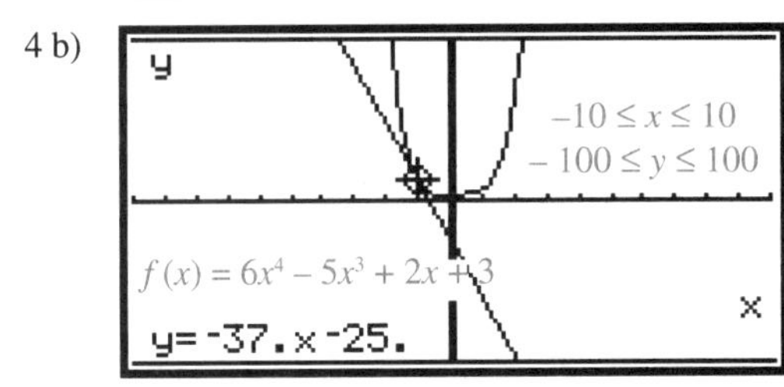

4 c)

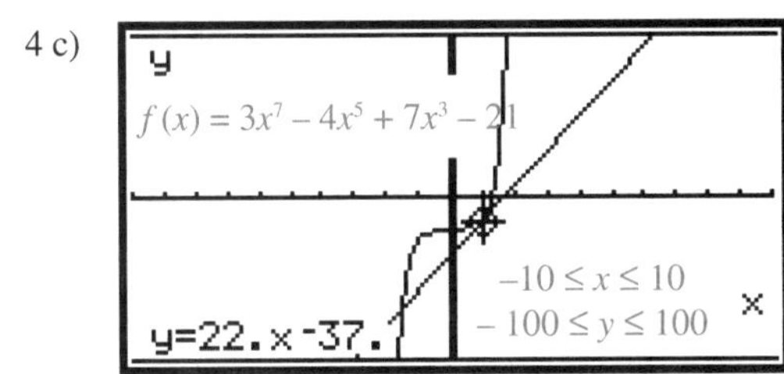

Exploration 8 (cont'd)

5. The graphs of $f(x)$ and $f'(x)$ are shown in the displays below.

5 a)

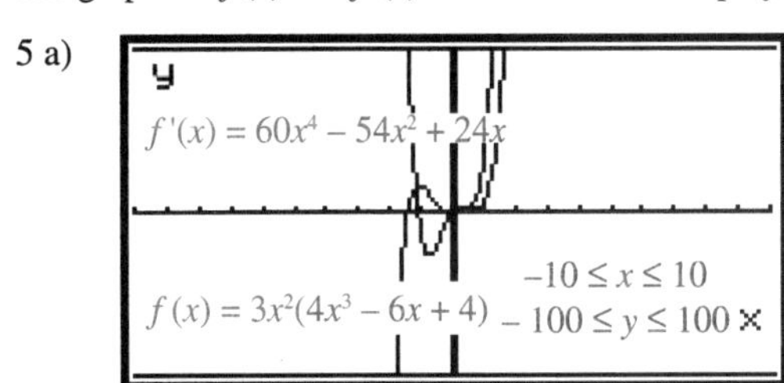

5 b)

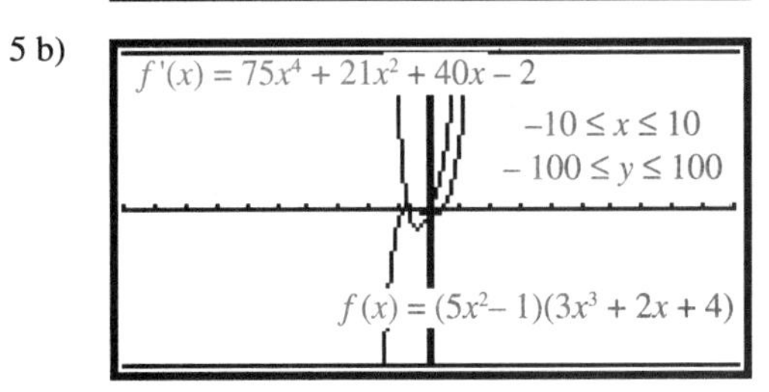

5 c)

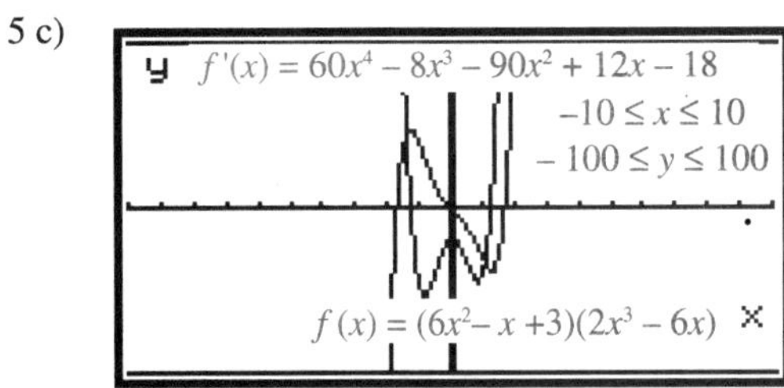

5 d)

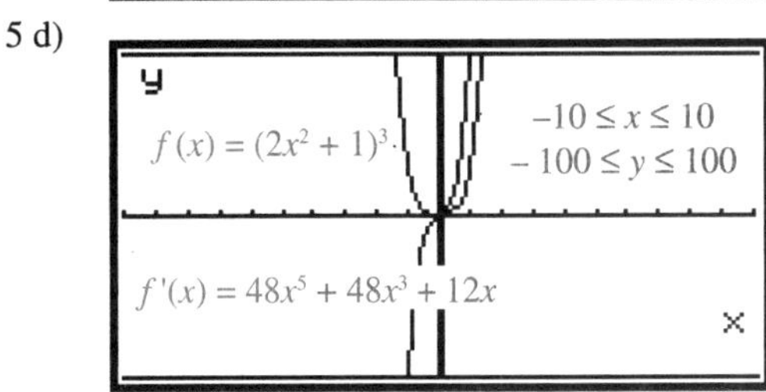

6. a) $y = -21.75x - 6$
   b) $y = 134x - 98$
   c) $y = 542x - 984$
   d) $y = 1$

7. a) $\dfrac{-3}{4x^2-7} - \dfrac{42}{(4x^2-7)^2}$
   b) $\dfrac{5}{2} - \dfrac{475}{2(2x-9)^2}$
   c) $\dfrac{1}{7}\left(\dfrac{1}{x^2} - \dfrac{51}{(3x-7)^2}\right)$

10. a) The instantaneous velocity of the VIPER at time t is given by: $v = -0.21t^2 + 6.3t + 1.2$
    b) The maximum velocity of 48.45 m/s occurs when $t = 15$ s.

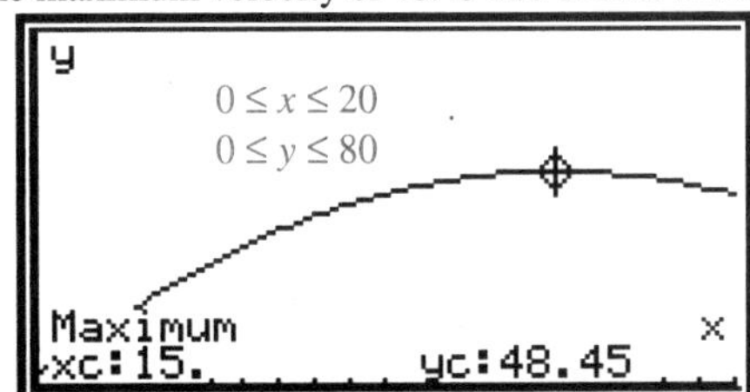

c) The acceleration at time t is given by $a = 6.3 - 0.42t$.
d) The instantaneous velocity of the VIPER at $t = 5.5$ is given by: $v = -0.21(5.5)^2 + 6.3(5.5) + 1.2$ or 29.4975 as found in exercise 9e).

Exploration 9

1. a) $y = \frac{1}{4}x^{-\frac{3}{4}}$ b) $y = -\frac{1}{2}x^{-\frac{3}{2}}$ c) $y = -x^{-\frac{2}{3}}$

2. The displays below show the commands and their outputs on the TI-89.

a)

$\blacksquare \frac{d}{dx}\left(x^{1/3} + 3\cdot x^{1/4} - x^{-2}\right)$ $\quad \frac{1}{3\cdot x^{2/3}} + \frac{3}{4\cdot x^{3/4}} + \frac{2}{x^3}$

`...x^(1/3)+3x^(1/4)-x^-2,x)`

b)

$\blacksquare \frac{d}{dx}\left(x^{-1/2} + x^{1/2}\right)$ $\quad \frac{1}{2\cdot\sqrt{x}} - \frac{1}{2\cdot x^{3/2}}$

`d(x^-(1/2)+x^(1/2),x)`

c)

$\blacksquare \frac{d}{dx}\left((3+\sqrt{x})^2\right)$ $\quad \frac{\sqrt{x}+3}{\sqrt{x}}$

`d((3+√(x))^2,x)`

d)

$\blacksquare \frac{d}{dx}\left(\left(\frac{1}{\sqrt{x}} + \sqrt{x}\right)^2\right)$ $\quad \frac{(x-1)\cdot(x+1)}{x^2}$

`d((1/√(x)+√(x))^2,x)`

3. We cannot apply the simple power rule directly because

$$\frac{d}{d(1+2x)}\left((1+2x)^{\frac{1}{2}}\right) = \frac{1}{2}(1+2x)^{-\frac{1}{2}}, \text{ but } \frac{d}{dx}\left((1+2x)^{\frac{1}{2}}\right)$$

requires an additional factor of $\frac{d}{dx}(1+2x)$.

Note: It is a very common error for students to apply the simple power rule when the generalized rule is needed.

4. a)

$\blacksquare \frac{d}{dx}\left(\sqrt{5+3\cdot x}\right)$ $\quad \frac{3}{2\cdot\sqrt{3\cdot x+5}}$

`d(√(5+3x),x)`

b)

$\blacksquare \frac{d}{dx}\left(\sqrt{6\cdot x^2 - 3\cdot x + 1}\right)$ $\quad \frac{3\cdot(4\cdot x - 1)}{2\cdot\sqrt{6\cdot x^2 - 3\cdot x + 1}}$

`d(√(6x^2-3x+1),x)`

c)

$\blacksquare \frac{d}{dx}\left((5\cdot x^2 - 7\cdot x)^{1/3}\right)$ $\quad \frac{(5\cdot x - 7)^{1/3}}{3\cdot x^{2/3}} + \frac{5\cdot x^{1/3}}{3\cdot(5\cdot x - 7)^{2/3}}$

`d((5x^2-7x)^(1/3),x)`

d)

$\blacksquare \frac{d}{dx}\left(\frac{1}{\sqrt{6\cdot x^2 - 3\cdot x + 1}}\right)$ $\quad \frac{-3\cdot(4\cdot x - 1)}{2\cdot(6\cdot x^2 - 3\cdot x + 1)^{3/2}}$

`d(1/√(6x^2-3x+1),x)`

Exploration 9 (cont'd)

5. The displays below show the commands and their TI-89 outputs.

a)

$\blacksquare \frac{d}{dx}\left((3\cdot x + 5)^4\right)$ $\quad 12\cdot(3\cdot x + 5)^3$

`d((3x+5)^4,x)`

b)

$\blacksquare \frac{d}{dx}\left(\sqrt{2\cdot x - 7}\right)$ $\quad \frac{1}{\sqrt{2\cdot x - 7}}$

`d(√(2x-7),x)`

c)

$\blacksquare \frac{d}{dx}\left((2\cdot x - 1)^3\cdot(3\cdot x + 4)^2\right)$ $\quad 6\cdot(2\cdot x - 1)^2\cdot(3\cdot x + 4)\cdot(5\cdot x + 3)$

`d((2x-1)^3(3x+4)^2,x)`

d)

$\blacksquare \frac{d}{dx}\left((5+3\cdot x)^{-1/2}\right)$ $\quad \frac{-3}{2\cdot(3\cdot x + 5)^{3/2}}$

`d((5+3x)^(-1/2),x)`

e)

$\blacksquare \frac{d}{dx}\left(\frac{(2\cdot x - 1)^3}{3\cdot x + 4}\right)$ $\quad \frac{3\cdot(2\cdot x - 1)^2\cdot(4\cdot x + 9)}{(3\cdot x + 4)^2}$

`d((2x-1)^3/(3x+4),x)`

f)

$\blacksquare \frac{d}{dx}\left(\sqrt{\frac{3\cdot x^2 + 1}{5\cdot x - 2}}\right)$ $\quad \frac{(15\cdot x^2 - 12\cdot x - 5)\cdot\left(\frac{1}{5\cdot x - 2}\right.}{2\cdot\sqrt{3\cdot x^2 + 1}}$ ▸

`d(√((3x^2+1)/(5x-2)),x)`

6 a) $x = 3/2$ or $x = 1/3$ b) $x = -3$

c) $x = 16/3$ or 0 d) $x = 0$ or $2^{-\frac{1}{3}}$ or $0.05549\ldots$

e) $x = \frac{1+\sqrt{97}}{8}$ or $\frac{1-\sqrt{97}}{8}$ f) $x = \frac{6+\sqrt{111}}{15}$ or $\frac{6-\sqrt{111}}{15}$

8. BC + CF = BC + CF*, so the distance BC + CF which we want to minimize, is the same as the distance from B to F* through C. But we know that the distance from B to F* through C is a minimum when BCF* is a straight line, i.e. C is the intersection of the line BF* with the cart path. C is the intersection of the line segment joining (0, 6) to (15, –10) with the x-axis.

11. a) $V = \frac{1}{3}\pi r^2(x+16)$ b) $PS = \sqrt{x^2 - 16^2}$

c) Both are right triangles with angle QPR in common, so they are equiangular and hence similar.

d) $\frac{16}{\sqrt{x^2-16^2}} = \frac{r}{x+16}$ e) $r = \frac{16(x+16)}{\sqrt{x^2-16^2}}$

f) $V = \frac{1}{3}\pi 16^2 \frac{(x+16)^3}{x^2-16^2} = 16^2\left(\frac{\pi}{3}\right)\frac{(x+16)^2}{x-16}$

g) $x = 48$ cm. $r = 16\sqrt{2}$ and $V = 32768\pi/3$.

Exploration 10

1. a) e^x is defined for all real values of x.
   b) e^x is not negative for any real value of x.
   c) e^x has no local extrema because it is a monotonically increasing function.

2. a) The graphs of e^x and e^{-x} are reflections of each other in the y-axis.
   b) The slope of the tangent to the graph of e^x at $x = x_0$ is the negative of the slope of the tangent to the graph of e^{-x} at $x = x_0$.

3. a) (i) $e^2 \approx 7.389$ (ii) $e^3 \approx 20.085$ (iii) $e^{3.5} \approx 33.115$
   b) The tangent line at $x = 3.5$ is shown in this display.

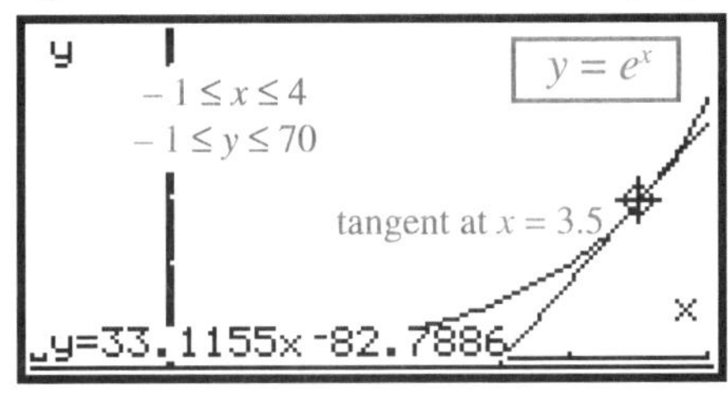

4. The displays below show the commands and their outputs on the TI-89.

a)

$$\frac{d}{dx}\left(e^{3\cdot x^2+5\cdot x+1}\right)$$
$$(6\cdot e\cdot x+5\cdot e)\cdot e^{3\cdot x^2+5\cdot x}$$
```
d(e^(3x^2+5x+1),x)
```

b)

$$\frac{d}{dx}\left(e^{\frac{-1}{x}}\right) \qquad \frac{e^{\frac{-1}{x}}}{x^2}$$
```
d(e^(-1/x),x)
```

c)

$$\frac{d}{dx}\left(e^{\sqrt{3\cdot x^2+5}}\right)$$
$$\frac{3\cdot x\cdot e^{\sqrt{3\cdot x^2+5}}}{\sqrt{3\cdot x^2+5}}$$
```
d(e^(√(3x^2+5)),x)
```

d)

$$\frac{d}{dx}\left(\frac{1+e^{\wedge}}{1-e^{-x}}\right)$$
$$\frac{e^{2\cdot x}-2\cdot e^x-1}{4\cdot\left(\sinh\left(\frac{x}{2}\right)\right)^2}$$
```
d((1+e^(x))/(1-e^(-x)),x)
```

or equivalently,

$$\frac{d}{dx}\left(\frac{1+e^x}{1-e^{-x}}\right) \qquad \frac{e^x\cdot\left(e^{2\cdot x}-2\cdot e^x-1\right)}{\left(e^x-1\right)^2}$$
```
d((1+e^(x))/(1-e^(-x)),x)
```

5. a) $y = 25.86x + 2.03$ b) $y = 0.0045x - 0.0004$
   c) $y = 1.26x + 9.29$ d) $y = -198.72x + 42$

Exploration 10 (cont'd)

6. The displays below show the commands and their outputs on the TI-89.

Note: The third entry in the ***d*** (command indicates the order of the derivative and that its default value is 1.

a)

$$\frac{d^2}{dx^2}\left(e^x\right) \qquad e^x$$
```
d(e^(x),x,2)
```

b)

$$\frac{d^2}{dx^2}\left(e^{-2\cdot x}\right) \qquad 4\cdot e^{-2\cdot x}$$
```
d(e^(-2x),x,2)
```

c)

$$\frac{d^2}{dx^2}\left(x\cdot e^x\right) \qquad (x+2)\cdot e^x$$
```
d(xe^(x),x,2)
```

7. The TI-89 command and the output are shown in this display.

$$\text{solve}\left(e^x - e^{-x} = 1, x\right)$$
$$x = \ln\left(\frac{\sqrt{5}+1}{2}\right)$$
```
solve(e^(x)-e^(-x)=1,x)
```

To solve this equation *without* the TI-89, multiply both sides of the equation by e^x and observe that since $e^{2x} = (e^x)^2$, we have a quadratic equation in e^x. This equation yields the solution $e^x = (1 + \sqrt{5})/2$. On taking logarithms of both sides, we obtain $x = \ln[(1+\sqrt{5})/2]$.

8. b) The displays show the commands and the solutions.

$$\text{solve}\left(\frac{d}{dx}\left(x\cdot e^{-x^2}\right) = 0, x\right)$$
$$x = \frac{\sqrt{2}}{2} \text{ or } x = \frac{-\sqrt{2}}{2}$$
```
solve(d(xe^(-x^2),x)=0,x)
```

$$\text{solve}\left(\frac{d}{dx}\left(x^2\cdot e^{-x}\right) = 0, x\right)$$
$$x = 2 \text{ or } x = 0$$
```
solve(d(x^2e^(-x),x)=0,x)
```

11. Your TI-89 yields the display below.

$$\frac{d}{dx}\left(\frac{78.12}{6.3+102\cdot e^{-.02817\cdot x}}\right)$$
$$\frac{5.65546\cdot(1.02857)^x}{\left((1.02857)^x+16.1905\right)^2}$$
```
.../(6.3+102e^(-.02817x)),x...
```

Substituting 1.02857^x for $e^{0.02817x}$ in the *Worked Example* and simplifying the resulting expression reduces it to the display above.

Selected Solutions to Exercises & Hints for the Investigations

Exploration 11

1. Since $\ln(x)$ is defined to be the inverse of e^x, then the application of either function followed by the other leaves x unchanged. That is, $e^{\ln(x)} = x$ and $\ln(e^x) = x$.

2. a) The natural logarithm is defined for $x > 0$.
   b) $\ln(x)$ is a monotonically increasing function and $\ln(e) = 1$, so $\ln x < 1$ when $x < e$ and $\ln x > 1$ when $x > e$.

3. Since $\frac{d}{dx}(\ln x) = \frac{1}{x}$, then
   a) slope < 1 for $x > 1$ b) slope > 1 for $x < 1$
   c) slope = 1 for $x = 1$ d) slope = 2 for $x = 1/2$

4. We must determine z where $z = \log_a y$. That is, $a^z = y$. Since it is given that $y = \ln x$, then $a^z = \ln x$. i.e. $(e^{\ln a})^z = \ln x$, so $e^{z\ln a} = \ln x$. Taking the natural log of both sides of this equation yields $z = \frac{\ln(\ln(x))}{\ln a}$.

5. The displays show the TI-89 commands and outputs.

a)
$$\frac{d}{dx}\left(2^{-x^2}\right)$$
$$-2\cdot\ln(2)\cdot x\cdot 2^{-x^2}$$
```
d(2^-x^2,x)
```

b)
$$\frac{d}{dx}\left(x^{x^2}\right)$$
$$(2\cdot x\cdot\ln(x)+x)\cdot x^{x^2}$$
```
d(x^(x^2),x)
```

c)
$$\frac{d}{dx}\left((2\cdot x+1)^x\right)$$
$$\left(\ln(2\cdot x+1)+\frac{2\cdot x}{2\cdot x+1}\right)\cdot(2\cdot x+1)^x$$
```
d((2x+1)^x,x)
```

6. The TI-89 command line and output are shown here.

$$\frac{-2\cdot\ln(3)\cdot x\cdot 3^{\frac{1}{x^2+1}}}{3\cdot(x^2+1)^2}$$
```
d(3^(-x^2/(1+x^2)),x)
```

To convert this output so that it has the same form as the Worked Example, write the 3 in the denominator of the above display as 3^{-1} in the numerator and combine it with the other power of 3 in the numerator.

7. a) $\frac{6x}{3x^2+2}$ b) $\frac{x}{x^2-3}$ c) $\frac{-x}{\ln 2\left(x^2+1\right)}$

8. b)
   i) $\sqrt{\frac{x^2+1}{x^2+4}}\left(\frac{3x}{(x^2+1)(x^2+4)}\right)$ ii) $\frac{5}{3}\left(\frac{x^{-\frac{2}{3}}}{(3x+5)^{\frac{4}{3}}}\right)$

Exploration 11 (cont'd)

8 b)
   iii) $\left(\frac{-x^2+6x+5}{(x^2+5)^{\frac{3}{2}}\sqrt{x^2+2x-1}}\right)$

9. a) $\frac{d}{dx}\left(\frac{1}{1+\ln x}\right) = \frac{d}{dx}(1+\ln x)^{-1} = -(1+\ln x)^{-2}\frac{d}{dx}(1+\ln x)$
$$= \frac{-1}{x(1+\ln x)^2}$$

b) $\frac{d}{dx}\left(\frac{\ln x}{x}\right) = \ln x\frac{d}{dx}x^{-1} + x^{-1}\frac{d}{dx}(\ln x) = \frac{(1-\ln x)}{x^2}$

c)
$$\frac{d}{dx}\left(\ln\sqrt{\frac{(1+x)(1+2x)(1+3x)}{(1-x)(1-2x)(1-3x)}}\right) = \frac{1}{2}\frac{d}{dx}\left(\ln\left(\frac{(1+x)(1+2x)(1+3x)}{(1-x)(1-2x)(1-3x)}\right)\right)$$
$$= \frac{1}{2}\frac{d}{dx}\left[\ln(1+x)+\ln(1+2x)+\ln(1+3x)-\ln(1-x)+\ln(1-2x)+\ln(1-3x)\right]$$
$$= \frac{1}{2}\left[\frac{1}{(1+x)}+\frac{2}{(1+2x)}+\frac{3}{(1+3x)}+\frac{1}{(1-x)}+\frac{2}{(1-2x)}+\frac{3}{(1-3x)}\right]$$
$$= \left[\frac{1}{(1-x^2)}+\frac{2}{(1-4x^2)}+\frac{3}{(1-9x^2)}\right]$$

11. a) $\left(\frac{1}{\ln(10)}\right)\frac{6x}{3x^2+2}$ b) $\left(\frac{1}{\ln(10)}\right)\frac{x}{x^2-3}$
    c) $\left(\frac{1}{\ln(10)}\right)\left[\frac{6x}{3x^2+2}-\frac{x}{x^2-3}\right]$ d) negative of c)

How Many Prime Numbers up to N?

We observe that the function $P(x)$ is not a good approximation to $\pi(x)$ for small values of x, but as x increases in value, $P(x)$ becomes a closer approximation than $p(x)$. Note that it takes a long time to graph $P(x)$ for large x, but $p(x)$ plots very quickly. Why do you think this is so? How close is $p(1000)$ to the true value $\pi(1000) = 168$? Does the difference between P(x) and $p(x)$ increase or decrease as x increases? Does their ratio increase or decrease?

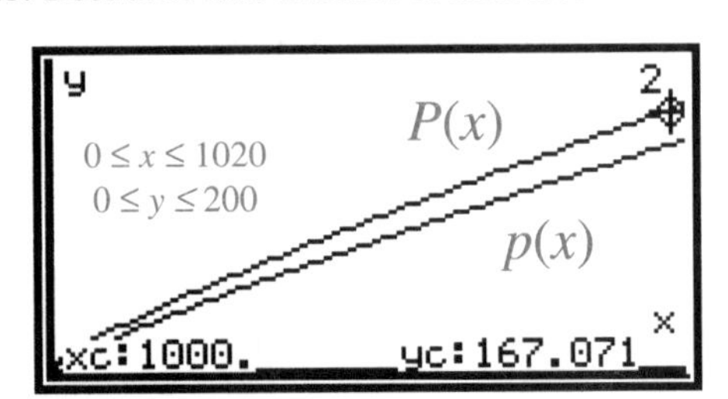

Exploration 12

2. We know from the graph of $y = \sin x$, that its maxima occur at $x = \pi/2 \pm 2n\pi$ and that its minima occur at $x = 3\pi/2 \pm 2n\pi$, where n is an integer. Therefore, we need only set the argument equal to these values and solve for x.

3. Application of the formulas for the derivative of the generalized sine and cosine functions yield these derivatives.
a) $2\sin x \cos x$ or $\sin 2x$ b) $3\cos(3x + 5)$ c) $-5\sin(5x + 1)$

4. We first express the reciprocal trig ratios in terms of sine and cosine and then differentiate. We must also do this on the TI-89.

a) $$\frac{d}{dx}(\sec x) = \frac{d}{dx}\left((\cos x)^{-1}\right) = -(\cos x)^{-2}(-\sin x) = \sec x \tan x$$

b) $$\frac{d}{dx}(\tan x) = \frac{d}{dx}\left(\frac{\sin x}{\cos x}\right) = \frac{\cos^2 x + \sin^2 x}{\cos^2 x} = \sec^2 x$$

c) $$\frac{d}{dx}(\csc(3x+1)) = \frac{d}{dx}\left((\sin(3x+1))^{-1}\right)$$
$$= -3(\sin(3x+1))^{-2}(\cos(3x+1))$$
$$= 3\csc(3x+1)\cot(3x+1)$$

5. To convert the argument to radians, we must multiply by $\pi/180$, and we obtain
$$y = 3\sin\left(\frac{\pi}{180}(5x+6)\right) \text{ and so,}$$
$$\frac{dy}{dx} = 15\left(\frac{\pi}{180}\right)\cos\left(\frac{\pi}{180}(5x+6)\right) = \frac{\pi}{12}\cos\left(\frac{\pi}{180}(5x+6)\right)$$

6. a) $\frac{dx}{dt} = -\frac{8\pi}{3}\sin\left(\frac{2\pi t}{3}\right)$ and $\frac{d^2x}{dt^2} = -\frac{16\pi^2}{9}\cos\left(\frac{2\pi t}{3}\right)$

b) Displacement is the *absolute value* of x, so the maximum displacement occurs whenever $\cos(2\pi t/3) = 1$ or -1. That is, the maxima occur whenever $2\pi t/3$ is a multiple of π; i.e. at $t = 3n/2$ where n is an integer.
Minima occur when $2\pi t/3 = (2n + 1)\pi/2$; i.e. $t = (6n + 3)/4$ where n is an integer.

c) Maximum velocity occurs when $2\pi t/3 = (2n + 1)\pi/2$; i.e. $t = (6n + 3)/4$ where n is an integer. That is, the pendulum reaches maximum velocity when its displacement is zero.

d) Maximum acceleration occurs whenever $\cos(2\pi t/3) = 1$ or -1. That is, maximum accelerations occurs when $t = 3n/2$ where n is an integer (i.e. when the velocity is zero).

7. The TI-89 yields $$\frac{14\cdot\pi\cdot\sin\left(\frac{2\cdot\pi\cdot x}{365} + \frac{9\cdot\pi}{146}\right)}{1095}$$

$$\sin\left(\frac{2\pi x}{365} + \frac{9\pi}{146}\right) = \cos\left(\frac{2\pi x}{365} + \frac{9\pi}{146} - \frac{\pi}{2}\right) \quad \sin A = \cos(A - \pi/2)$$
$$= \cos\left(\frac{2\pi}{365}(x - 80)\right)$$ The answers are the same.

Exploration 12 (cont'd)

8. Since $\frac{dy}{dx} = 0$ the graph of y has 0 slope everywhere, then y is constant. Substituting $x = 0$, reveals that the constant is 1.

9. Application of the generalized exponent rule yields

a) $\frac{dy}{dx} = \cos x\, e^{\sin x}$ $\frac{d^2y}{dx^2} = \left(\cos^2 x - \sin x\right)e^{\sin x}$

b) $\frac{dy}{dx} = (x\cos x + 1)e^{\sin x}$

$\frac{d^2y}{dx^2} = \left(x\cos^2 x - x\sin x + 2\cos x\right)e^{\sin x}$

c) The display shows the two derivatives. Note how we use

2nd [ANS]

to compute the second derivative. This technique is useful when computing higher order derivatives.

```
■ d/dx(ln(cos(x)))          -tan(x)
■ d/dx(-tan(x))             -1/(cos(x))^2
d(ans(1),x)
```

12. The general procedure is given below.

$$\frac{d}{dx}(\sin f(x))$$
$$= \lim_{\Delta x \to 0} \frac{\sin(f(x+\Delta x)) - \sin(f(x))}{\Delta x}$$
$$= \lim_{\Delta x \to 0} \frac{2\cos\left(\frac{f(x+\Delta x)+f(x)}{2}\right)\sin\left(\frac{f(x+\Delta x)-f(x)}{2}\right)}{\Delta x}$$
$$= \cos(f(x)) \lim_{\Delta x \to 0} \frac{2\sin\left(\frac{f(x+\Delta x)-f(x)}{2}\right)}{\Delta x}$$
$$= \cos(f(x)) \lim_{\Delta x \to 0} \frac{\sin\left(\frac{f(x+\Delta x)-f(x)}{2}\right)}{\left(\frac{f(x+\Delta x)-f(x)}{2}\right)} \times \frac{\left(\frac{f(x+\Delta x)-f(x)}{2}\right)}{\left(\frac{\Delta x}{2}\right)}$$
$$= \cos(f(x)) \lim_{\Delta x \to 0} \frac{\left(\frac{f(x+\Delta x)-f(x)}{2}\right)}{\left(\frac{\Delta x}{2}\right)}$$
$$= \cos(f(x))\frac{d}{dx}(f(x))$$

13. a) $|PQ| = f'(0)\Delta x$

d) L'Hôpital's rule states that if $f(0) = g(0) = 0$, then
$$\lim_{x\to 0}\frac{f(x)}{g(x)} = \lim_{x\to 0}\frac{f'(x)}{g'(x)}$$

e) (i) 1 (ii) $-2/\pi$ (iii) 0

SELECTED SOLUTIONS TO EXERCISES & HINTS FOR THE INVESTIGATIONS

Exploration 13

2. a) $\dfrac{6(3x+2)}{x(3x+4)}$ b) $\dfrac{-10(5x-1)\cos(5x-1)^2}{\sin^2(5x-1)^2}$

3. The diameter is decreasing at the rate of $\dfrac{25}{36\pi}$ cm/s.

4. If x is the distance of the base of the ladder from the wall and y denotes the height of the top of the ladder above the ground, then from the theorem of pythagoras,

$$y = \sqrt{7^2 - x^2} \text{ and so } \frac{dy}{dt} = \frac{dy}{dx}\frac{dx}{dt} = -\frac{x}{\sqrt{7^2 - x^2}}\frac{dx}{dt}$$

When $x = 4$, $dx/dt = 50$ *cm/s* $= 0.5$ *m/s*. Substitution into the above expression for dy/dt yields $\dfrac{dy}{dt} = \dfrac{4}{\sqrt{33}}$ *m/s* ≈ 69.6 *cm/s*

5. Proceeding as in exercise 4, we have

$$x = \sqrt{6^2 - y^2} \text{ and so } \frac{dx}{dt} = \frac{dx}{dy}\frac{dy}{dt} = -\frac{y}{\sqrt{6^2 - y^2}}\frac{dy}{dt}$$

When $y = 3$, $dy/dt = 10$ *cm/s* $= 0.1$ *m/s*. Substitution into the above expression for dx/dt yields $\dfrac{dx}{dt} = \dfrac{0.1}{\sqrt{3}}$ *m/s* ≈ 5.8 *cm/s*

6. Let x denote the aircraft's distance due east of the searchlight. Then $x = 1000\tan\theta$. Differentiating both sides of this equation with respect to the time t yields $\dfrac{dx}{dt} = 1000\sec^2\theta\dfrac{d\theta}{dt}$. Substituting $dx/dt = -150$ *m/s* and $\sec\theta = \sqrt{5}/2$ yields $-3/25$ radians/s.

7. Let x and y denote respectively the distance of the H.M.S. Coxeter and the S.S. Geodesic from the origin at time t (shown in the diagram) Then $x = 50 + 20t$ and $y = 40 - 30t$.
The distance, z between the two ships is given by $z = \sqrt{x^2 + y^2}$

i.e. $z = 10\sqrt{(5+2t)^2 + (4-3t)^2}$ and so $\dfrac{dz}{dt} = \dfrac{10(13t-2)}{\sqrt{13t^2 - 4t + 41}}$

a) Substituting $t = 1$ into the formula for dz/dt yields $11\sqrt{2}$ km/h. That is, after one hour the ships are departing at about 15.6 km/h.
b) The ships are closest when $dz/dt = 0$, i.e. when $t = 2/13$. At this time the ships are $230/\sqrt{13} \approx 63.8$ km apart.

8. The volume of the cylinder is given by $V = \pi r^2 h$. Also
$r^2 = R^2 - \left(\dfrac{h}{2}\right)^2$ and so, substitution yields $V = \pi\left(R^2 - \left(\dfrac{h}{2}\right)^2\right)h$

Differentiation yields $\dfrac{dV}{dh} = \pi R^2 - \dfrac{3}{4}\pi h^2$.

Solving $\dfrac{dV}{dh} = 0$ yields $h = \dfrac{2R}{\sqrt{3}}$.

11. e) The display verifies that $dy/dx = -0.311$ at $t = 1.2$

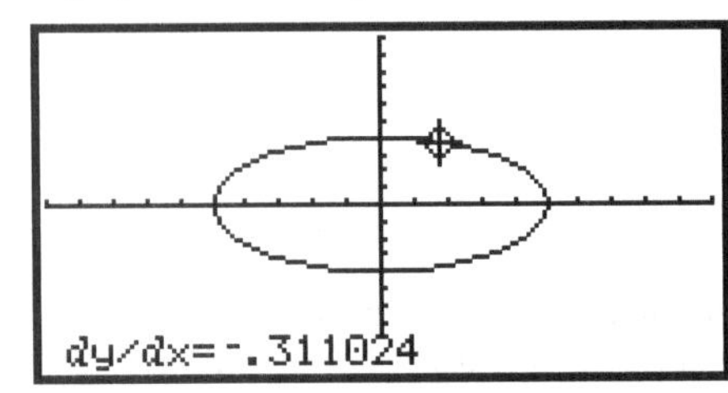

Exploration 14

1. $f(t) = \sqrt{t}$ is a continuous function on $[0, x]$, so $\int_0^x \sqrt{t}\,dt$ is differentiable and $\dfrac{d}{dx}\left(\int_0^x \sqrt{t}\,dt\right) = \sqrt{x}$.
Also from the power rule for derivatives, we know, $\dfrac{d\,x^{\frac{3}{2}}}{dx} = \dfrac{3}{2}x^{\frac{1}{2}}$
If we define $F(x) = \dfrac{2}{3}x^{\frac{3}{2}}$, then $\dfrac{d\,F(x)}{dx} = \sqrt{x}$. Therefore from the Corollary of the Fundamental Theorem of Calculus,

$$\int_0^x \sqrt{t}\,dt = F(x) - F(0) = \frac{2}{3}x^{\frac{3}{2}}$$

2. a) $\displaystyle\int_{-1}^{1}(x^3 - x)\,dx = \int_{-1}^{1}x^3\,dx - \int_{-1}^{1}x\,dx$

$$= \frac{1}{4}x^4\Big|_{-1}^{1} - \frac{1}{2}x^2\Big|_{-1}^{1} = 0$$

b) The lower left corner of the display indicates that the shaded signed area is zero.

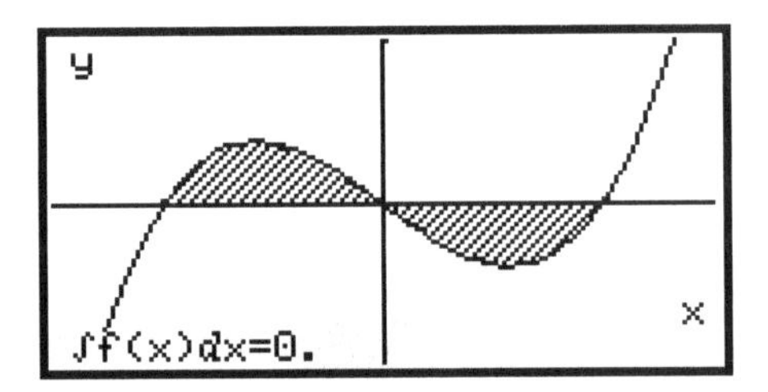

The area below the x-axis is regarded as "negative" area. Since the areas above and below the graph are equal in magnitude and opposite in sign, they add to zero.

c) $\displaystyle\int_{-1}^{0}(x^3 - x)\,dx = 0.25$ and $\displaystyle\int_{0}^{1}(x^3 - x)\,dx = -0.25$

d) Signed area is the net area when the (absolute) areas below the x-axis are subtracted from areas above the x-axis.

4. a) In each case, we apply the power rule for integration.

(i) $\displaystyle\int 3x^2\,dx = 3\left(\frac{x^3}{3}\right) + C = x^3 + C$

(ii) $\displaystyle\int (x+4)^{\frac{3}{2}}\,dx = \frac{2}{5}(x+4)^{\frac{5}{2}} + C$

(iii) $\displaystyle\int_1^5 \frac{1}{(x+5)^2}\,dx = \int_1^5 (x+5)^{-2}\,dx = -\frac{1}{(x+5)} + C$

b) & c) The displays show the TI-89 commands and outputs for (iii)

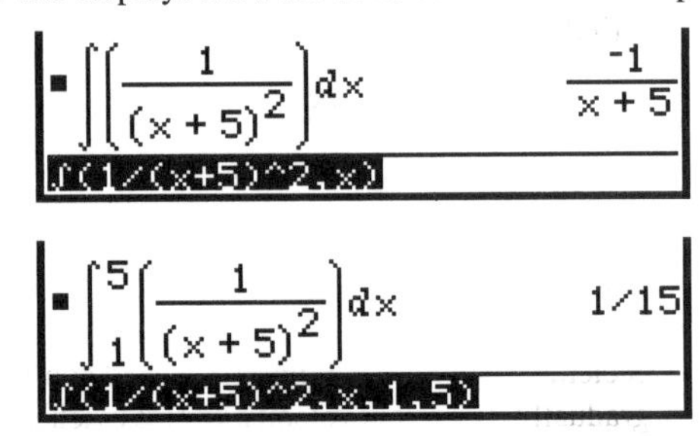

Exploration 14 (cont'd)

5. $$\int \left(\frac{x^3+x+1}{x^3}\right)dx = \int \left(\frac{x^3}{x^3}\right)dx + \int \left(\frac{x}{x^3}\right)dx + \int \left(\frac{1}{x^3}\right)dx$$

$$= x - \frac{1}{x} - \frac{1}{2x^2} + C$$

$$\int_{-1}^{1}\left(\frac{x^3+x+1}{x^3}\right)dx = \left[x - \frac{1}{x} - \frac{1}{2x^2}\right]_{-1}^{1} = 0$$

Note that your TI-89 indicates that the definite integral is undefined, however, our computation yields 0. Why does the integral appear to be undefined in the interval $-1 \le x \le 1$? Explain how the integral can have a finite value when the integrand has at least one infinite value.

6. $$\int (1-x)\sqrt{x}\,dx = \int \sqrt{x}\,dx - \int x^{\frac{3}{2}}\,dx = \frac{2}{3}x^{\frac{3}{2}} - \frac{2}{5}x^{\frac{5}{2}} + C$$

$$\int_0^6 (1-x)\sqrt{x}\,dx = \left[\frac{2}{3}x^{\frac{3}{2}} - \frac{2}{5}x^{\frac{5}{2}}\right]_0^6 = -\frac{52\sqrt{6}}{5}$$

8. b) The TI-89 commands and outputs are shown in the displays for (i) and (ii).

(i)

$$\blacksquare \int \left(\frac{2\cdot x+1}{(x^2+x+3)^2}\right)dx \qquad \frac{-1}{x^2+x+3}$$

```
∫((2x+1)/(x^2+x+3)^2,x)
```

(ii)

$$\blacksquare \int \left(\sqrt{x^2+5\cdot x}\cdot(2\cdot x+5)\right)dx \qquad \frac{2\cdot(x\cdot(x+5))^{3/2}}{3}$$

```
∫(√(x^2+5x)*(2x+5),x)
```

12. We use the fact that velocity is the derivative of distance with respect to time and acceleration is the derivative of velocity with respect to time. We are given that the acceleration is $-3\ m/s^2$, i.e.

$$\frac{d^2s}{dt^2} = \frac{d}{dt}\left(\frac{ds}{dt}\right) = -3\,m/s^2$$

Integration with respect to t yields $\frac{ds}{dt} = -3t + C$

Velocity when $t = 0$ is 60, so $C = 60$., therefore $\frac{ds}{dt} = -3t + 60$.

To obtain s, we integrate again with respect to t. This yields

$$s = -\frac{3}{2}t^2 + 60t + D$$

Substituting $t = 0$ into this expression, and using the fact that $s = 0$ when $t = 0$, we find $D = 0$, so

$$s = -\frac{3}{2}t^2 + 60t$$

Velocity is zero when $-3t + 60 = 0$, i.e. $t = -20$. Substituting $t = -20$ into the expression for s yields $s = 600\ m$.
That is, the projectile has traveled 600 m horizontally when its velocity has receded to zero. We note that a negative acceleration is actually a deceleration. The projectile is launched at 60 m/s and air resistance gradually erodes the velocity until it reaches zero.

Exploration 15

3. The TI-89 displays $-1/2\cos^2 x$.
Since $\cos 2x = 2\cos^2 x - 1 = 1 - 2\sin^2 x$ we can see that all these answers differ by a constant and therefore represent the same indefinite integral.

4. a) $$\int e^{-3x}\,dx = -\tfrac{1}{3}\int e^{-3x}\,d(\pm 3x) = -\tfrac{1}{3}e^{-3x} + C$$

b) $$\int 2^x\,dx = \frac{2^x}{\ln 2} + C$$

c) $$\int xe^{x^2}\,dx = \tfrac{1}{2}\int e^u\,du \text{ where } u = e^{x^2}$$

$$= \tfrac{1}{2}e^u + C = \tfrac{1}{2}e^{x^2} + C$$

5. Using the indefinite integrals we computed in exercise 4, we have

a) $$\int_0^1 e^{-3x}\,dx = -\frac{1}{3}e^{-3x}\Big|_0^1 = -\frac{1}{3}\left[e^{-3} - 1\right]$$

b) $$\int_0^5 2^x\,dx = \frac{2^x}{\ln 2}\Big|_0^5 = \frac{31}{\ln 2}$$

c) $$\int_1^2 xe^{x^2}\,dx = \frac{1}{2}e^{x^2}\Big|_1^2 = \frac{1}{2}e\left(e^3 - 1\right)$$

6. a) $2\ln x$ b) $$\int \frac{2x}{x^2+1}\,dx = \int \frac{1}{u}\,du \quad \text{where } u = x^2+1$$

$$= \ln u + C = \ln\left(x^2+1\right) + C$$

c) $$\int \frac{x+1}{2x^2+4x+3}\,dx = \frac{1}{4}\int \frac{4x+4}{2x^2+4x+3}\,dx$$

$$= \frac{1}{4}\int \frac{du}{u} \quad \text{where } u = 2x^2+4x+3$$

$$= \frac{1}{4}\ln\left(2x^2+4x+3\right) + C$$

7. a) $2\ln 2$ b) $2\ln 3$ c) $\frac{1}{4}\ln 73$

8. a) $$\int \frac{2}{x\ln x}\,dx = \int \frac{2}{u}\,du \quad \text{where } u = \ln x$$

$$= 2\ln u + C = 2\ln(\ln x) + C$$

b) $$\int \sin^2 x\cos x\,dx = \int u^2\,du \quad \text{where } u = \sin x$$

$$= \frac{u^3}{3} + C = \frac{\sin^3 x}{3} + C$$

c) $$\int \frac{\sin x}{\cos^3 x}\,dx = -\int \frac{du}{u^3} \quad \text{where } u = \cos x$$

$$= \frac{u^{-2}}{2} + C = \frac{1}{2\cos^2 x} + C$$

13. S(100) = 5.18737 S(1000) = 7.48547 S(5000) = 9.09450

14. a) There is no closed form for this integral. It must be evaluated using numerical approximations.
b) (i) 1 (ii) 0.5 (iii) 0.68269

Exploration 16

2. The output on the TI-89 follows directly from the answer in *Worked Example* 3, when the sums and differences of the logs are expressed as the logs of products and quotients respectively.

3. The TI-89 command and output are shown in the display below.

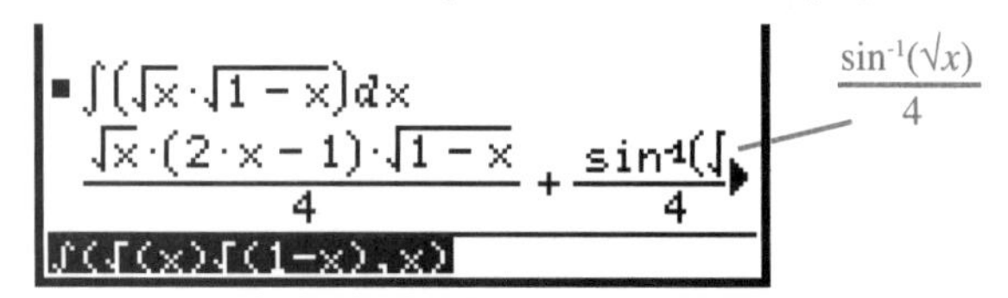

$\frac{\sin^{-1}(\sqrt{x})}{4}$

To reconcile the output of the TI-89 with the answer given on page 70, merely expand the product in the first term and simplify.

4. a) Let $u = x$ and $\frac{dv}{dx} = e^x$, then $du = dx$ and $v = e^x$

$$\int xe^x\,dx = xe^x - \int e^x\,dx = xe^x - e^x + C$$

b) Let $u = x$ and $\frac{dv}{dx} = \sqrt{x+3}$, then $du = dx$

and $v = \int \sqrt{x+3}\,dx = \frac{2}{3}(x+3)^{\frac{3}{2}}$

$$\int x\sqrt{x+3}\,dx = \frac{2}{3}(x+3)^{\frac{3}{2}} - \frac{2}{3}\int (x+3)^{\frac{3}{2}}\,dx$$

$$= \frac{2}{3}(x+3)^{\frac{3}{2}} - \frac{4}{15}(x+3)^{\frac{5}{2}} + C$$

c) Let $u = x^2$ and $\frac{dv}{dx} = \cos x$, then $du = 2x\,dx$ and $v = -\sin x$

$$\int x^2 \cos x\,dx = -x^2 \sin x + 2\int x \sin x\,dx$$

Then integrate $2\int x \sin x\,dx$ by parts to get

$$\int x^2 \cos x\,dx = -x^2 \sin x + 2x\cos x - 2\sin x + C$$

8. If you get stuck on any integration, compute the integral on your TI-89 and work backwards to discover the required substitution.

a)

```
■ ∫(x^3·√(1 - x^2))dx
              -(3·x^2 + 2)·(1 - x^2)^(3/2)
              ----------------------------
                          15
∫(x^3*(√(1-x^2)),x)
```

b)

```
■ ∫(x·√(1 + x))dx
              2·(x + 1)^(3/2)·(3·x - 2)
              -------------------------
                         15
∫(x√(1+x),x)
```

11. The area between $x = 0$ and $x = 1$ is 4/45.

12. The reduction formulas for the sine and cosine are

$$\int \sin^n x\,dx = -\frac{1}{n}\sin^{n-1} x \cos x + \frac{n-1}{n}\int \sin^{n-2} x\,dx$$

$$\int \cos^n x\,dx = \frac{1}{n}\cos^{n-1} x \sin x + \frac{n-1}{n}\int \cos^{n-2} x\,dx$$

Exploration 17

2. a) & b) The display below shows the graph of $y1(x)$ and the tangent line at $x = 2$.

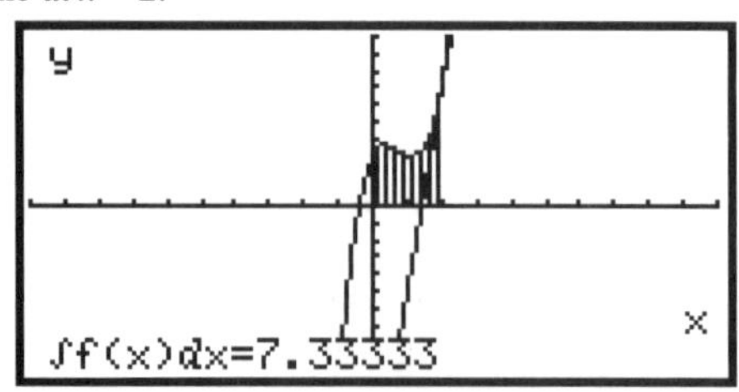

c) The shaded area in the display above is the area under the curve $y1(x)$ in the interval $0 \le x \le 2$. The area under the curve $y1(x)$ but left of the tangent line is given by the area under $y1(x)$ between $0 \le x \le 2$ less the area under the tangent line and in the interval $17/12 \le x \le 2$.

That is, $\int_0^2 (3x^3 - 7x^2 + 4x + 3)\,dx - \int_{\frac{17}{12}}^2 (12x - 17)\,dx$

d) $\left[3/4x^4 - 7/3x^3 + 2x^2 + 3x\right]\Big|_0^2 - \left[6x^2 - 17x\right]\Big|_{17/12}^2$

$= 22/3 - 49/24 = 127/24 \approx 5.291\ldots$

e) The split screen display shows that the difference in the areas is $7.3333\ldots - 2.0416\ldots \approx 5.291\ldots$

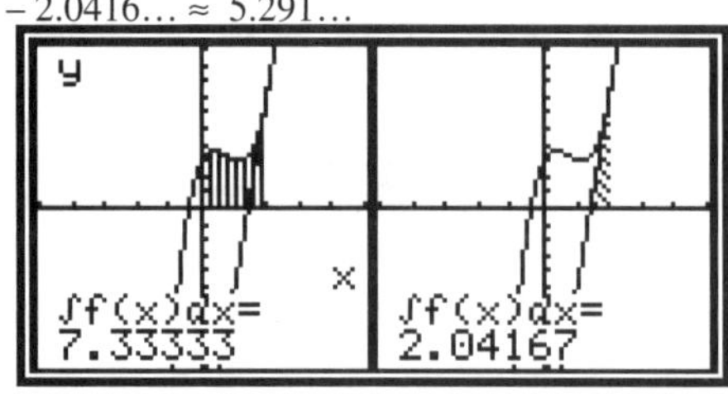

3 a)

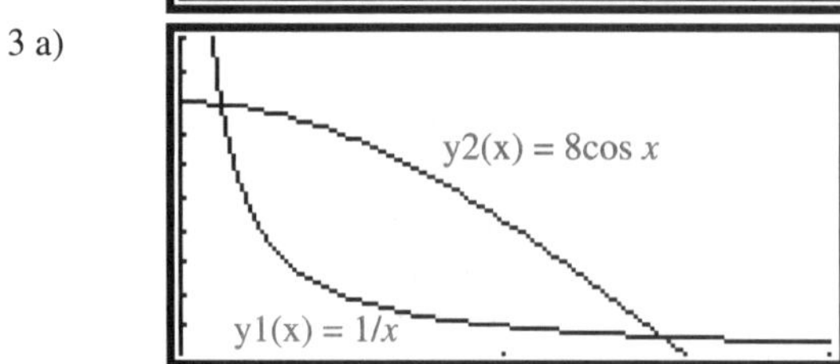

b) Proceeding as in exercise 2, we obtain the intersection points (0.12599…, 7.93658…) and (1.48661…, 0.67266…).

c) The definite integral of $y2(x) - y1(x)$ between these two intersection points is given by

$$\int_{0.126}^{1.49} \left(8\cos x - \frac{1}{x}\right) dx = \left[8\sin x - \ln x\right]\Big|_{0.126}^{1.49} = 4.498\ldots$$

d) To evaluate the definite integral in part c) using the integral command on the F3 menu, we enter the command shown in the display below. This output verifies our answer in part c).

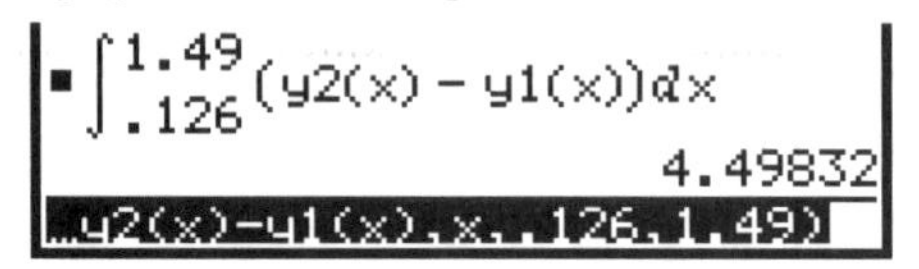

4. Application of the formula for arc length yields

$$S = \int_0^7 \sqrt{1+9x}\,dx = \frac{2}{27}\left[(1+9x)^{\frac{3}{2}}\right]\Bigg|_0^7 = \frac{2}{27}(511) \approx 37.851\ldots$$

The TI-89 Arc command from the graph screen verifies this value.

Exploration 17 (cont'd)

5. a) The arc length between the $x = 1/\sqrt{2}$ and $x = \sqrt{3}/2$, is given by

$$S=\int_{1/\sqrt{2}}^{\sqrt{3}/2} \frac{dx}{1-x^2} = \sin^{-1} x \Big|_{1/\sqrt{2}}^{\sqrt{3}/2} = \frac{\pi}{12}$$

b) These points subtend an angle of $\pi/4 - \pi/6$ or $\pi/12$ radians, so the arc on the unit circle is $\pi/12$ units long (by definition of a radian)

6. b) The point of intersection is (4, 5.656...).
   c) & d) The enclosed area of the enclosed region is given by

$$\int_0^4 \frac{1}{3}\frac{2x-8}{\sqrt{3x^2-24x+50}}\,dx = -\frac{2}{3}\left[\sqrt{3x^2-24x+50}\right]\Big|_0^4 = \frac{8\sqrt{2}}{3}$$

   e) The display below verifies the answer in part d).

$$\blacksquare \int_0^4 (y2(x) - y1(x))dx \qquad \frac{8\cdot\sqrt{2}}{3}$$

```
...y2(x)-y1(x),x,0,4)
```

8. b) The arc length between $x = -2.5$ and $x = 2.5$ is given by

$$\int_{-2.5}^{2.5} \sqrt{1+64x^2}\,dx = \frac{1}{16}\left[8x\sqrt{1+64x^2} + \ln\left|\sqrt{1+64x^2}+8x\right|\right]\Big|_{-2.5}^{2.5}$$

$$\approx 50.523\ldots$$

c) The display below confirms our result in part b).

$$\blacksquare \int_{-2.5}^{2.5} \sqrt{1 + 64\cdot x^2}\,dx \qquad 50.5236$$

```
∫(√(1+64x^2),x,-2.5,2.5)
```

9. The minimum value is attained when $\alpha = \frac{-1+\sqrt{5}}{2}$.

Exploration 18

1. We can reduce the paradox from 3 dimensions to 2 dimensions by realizing that a shape of finite area can have an infinite perimeter. So a finite amount of paint can cover an area of infinite perimeter.

2 a) Using the disk method, we find that the disk y units above the x-axis is a circle of radius $1 - y$ and thickness Δy. Therefore the area of the disk is $\pi(1 - y)^2$. Therefore the volume of the solid of revolution is

$$V = \int_0^1 \pi(1-y)^2\,dy = \pi\left[\frac{(y-1)^3}{3}\right]\Big|_0^1 = \frac{\pi}{3}$$

2 b) The solid of revolution is a cone with equation shown in the display below.

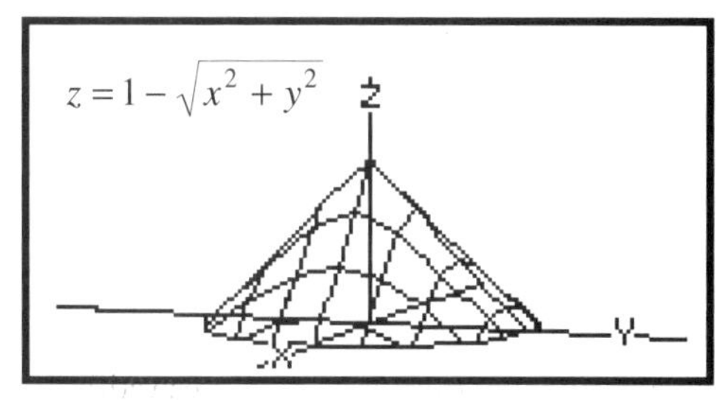

Exploration 18 (cont'd)

4. The disk method yields

$$V = \int_0^8 \pi y^2\,dx = \pi\int_0^8 9x\,dx = \left[\frac{9\pi x^2}{2}\right]\Big|_0^8 = 288\pi \text{ cubic units}$$

6. a) The graph of the hyperboloid is

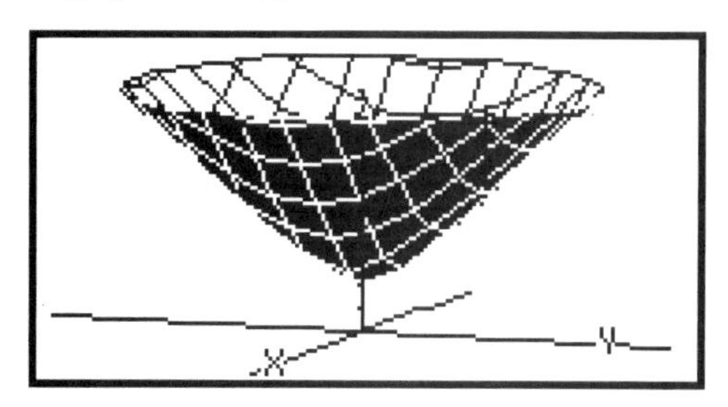

b) A slice parallel to the x-y plane and z_i units above it is an ellipse with equation $4x^2 + y^2 = 9z_i^2 - 36$, and area $(9z_i^2 - 36)\pi/2$.
The volume of the slice is $\Delta V_i = (9z_i^2 - 36)\pi/2\ \Delta z$.
The total volume is

$$V = \lim_{n\to\infty}\sum_{i=1}^{n}\Delta V_i = \lim_{n\to\infty}\sum_{i=1}^{n}\frac{\pi}{2}(9z_i^2-36)\,\Delta z$$

$$=\int_4^8 \frac{\pi}{2}(9z^2-36)dz = \frac{\pi}{2}\left[(3z^3-36z)\right]\Big|_4^8 = 600\pi$$

10. We set $u=\sqrt{x^4+1}$ so, $x^4 = u^2 - 1$, and $dx = \frac{u}{2x^3}$

$$\int_{1/b}^{1}\frac{\sqrt{x^4+1}}{x}\,dx = \int_{x=1/b}^{x=1}\left(\frac{u}{x}\right)\left(\frac{u}{2x^3}\right)du = \frac{1}{2}\int_{x=1/b}^{x=1}\left(\frac{u^2}{u^2-1}\right)du$$

$$= \frac{1}{2}\int_{x=1/b}^{x=1}\left(\frac{-1}{2(u+1)} + \frac{1}{2(u-1)} + 1\right)du$$

$$= \frac{1}{2}\int_{x=1/b}^{x=1}\frac{-1}{2(u+1)}\,du + \frac{1}{2}\int_{x=1/b}^{x=1}\frac{1}{2(u-1)}\,du + \frac{1}{2}\int_{x=1/b}^{x=1}du$$

$$= \left[-\frac{1}{4}\ln(u+1) + \frac{1}{4}\ln(u-1) + \frac{1}{2}u\right]\Bigg|_{x=1/b}^{x=1}$$

$$= \left[\frac{2\sqrt{x^4+1} + 2\ln\left(\sqrt{x^4+1}-1\right) - 4\ln|x|}{4}\right]\Bigg|_{1/b}^{b}$$

11. This is the solid of revolution we generated in *Worked Example* 1

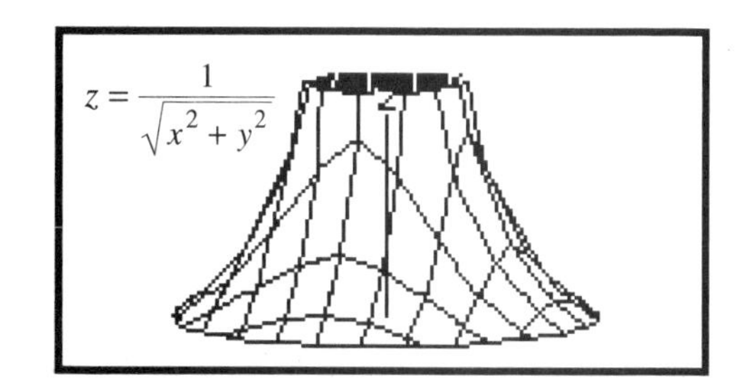